REASONING BY MATHEMATICAL INDUCTION IN CHILDREN'S ARITHMETIC

ADVANCES IN LEARNING AND INSTRUCTION SERIES

Series Editors:
S. Strauss, E. De Corte, R. Wegerif, K. Littleton

Further details: http://www.socscinet.com/education

Published

VAN SOMEREN, REIMANN, BOSHUIZEN & DE JONG
Learning with Multiple Representations

DILLENBOURG
Collaborative Learning: Cognitive and Computational Approaches

BLISS, SÄLJÖ & LIGHT
Learning Sites: Social and Technological Resources for Learning

KAYSER & VOSNIADOU
Modelling Changes in Understanding

SCHNOTZ, VOSNIADOU & CARRETERO
New Perspectives on Conceptual Change

KOZULIN & RAND
Experience of Mediated Learning

ROUET, LEVONEN & BIARDEAU
Multimedia Learning: Cognitive and Instructional Issues

GARRISON & ARCHER
A Transactional Perspective on Teaching and Learning

COWIE & AALSVOORT
Social Interaction in Learning and Instruction

VOLET & JÄRVELÄ
Motivation in Learning Contexts

Forthcoming Titles
BROMME & STAHL
Writing Hypertext and Learning: Conceptual and Empirical Approaches

Related journals — sample copies available online from:
http://www.elsevier.com or *http://www.socscinet.com/education*

Learning and Instruction
International Journal of Educational Research
Computers and Education
The Internet and Higher Education
Early Childhood Research Quarterly
Journal of Early Childhood Teacher Education
Learning and Individual Differences

REASONING BY MATHEMATICAL INDUCTION IN CHILDREN'S ARITHMETIC

LESLIE SMITH

Department of Educational Research, Lancaster University, UK

Published in Association with the European Association
for Learning and Instruction

2002

Pergamon
An imprint of Elsevier Science

Amsterdam – Boston – London – New York – Oxford – Paris
San Diego – San Francisco – Singapore – Sydney – Tokyo

ELSEVIER SCIENCE Ltd
The Boulevard, Langford Lane
Kidlington, Oxford OX5 1GB, UK

First edition 2002

Library of Congress Cataloging in Publication Data
A catalog record from the Library of Congress has been applied for.

British Library Cataloguing in Publication Data
A catalogue record from the British Library has been applied for.

ISBN: 0–08–044128–9

∞ The paper used in this publication meets the requirements of ANSI/NISO Z39.48–1992 (Permanence of Paper).
Printed in The Netherlands.

Contents

About the Authors

Leslie Smith
is Professor of Psychology and Epistemology of Development at Lancaster University. His research interests include:

- Piaget's theory and research
- developmental epistemology
- development of modal understanding
- developmental psychology applied to education.

Department of Educational Research
Lancaster University, Lancaster, UK

Damon Berridge
is Lecturer in Applied Statistics at Lancaster University. His research interests include:

- methods for duration and repeated ordered categorical data
- medical and social science applications.

Centre for Applied Statistics
Lancaster University, Lancaster, UK

Acknowledgements

My thanks to:

- the children and their teachers at Castle Park School, Dean Gibson RC School and St Thomas CE School in Kendal;
- Alice Jesmont and Katherine Beale for their commitment in transcribing the audio-recorded interviews;
- Angela Gelston for her enthusiasm and expertise in data processing;
- Robert Campbell for the reference to John Flavell's 1970 paper;
- colleagues in their responses to conference and seminar presentations of parts of this work, which included the British Psychological Society Developmental Psychology Section Annual Conference in Lancaster, the Department of Education at the University of Cyprus, the Department of Educational Research at Lancaster University, the Department of Psychology at the University of Edinburgh, and the Jean Piaget Society Annual Meeting in Chicago;
- Lancaster University for sabbatical leave in which this study was completed;
- The Leverhulme Trust in providing a personal research grant (RF&G/4/9700405) which made a welcome contribution to the funding of this project.

Finally, but most of all, I thank my wife Rose for her assistance and understanding throughout this work. Naturally enough, though, all the mistakes are mine.

Leslie Smith
April 2001

Summary

The central argument in this study is that reasoning by mathematical induction is under development during childhood. This argument is distinctive in four respects:

- its starting point is in first principles which were central to a dispute at the turn of the last century. Frege and Russell argued that all mathematical reasoning is logical. Poincaré denied this, insisting on the indispensability of intuition in mathematical reasoning. This dispute in the philosophy of mathematics is currently unresolved due to the inadequacy of both positions;
- its motivation is an important but neglected study. Forty years ago, Inhelder and Piaget identified a testable mechanism of iterative action underpinning reasoning by induction in mathematics which they interpreted by analogy with Poincaré's position contrary to the Frege–Russell position. Their evidence led to two hypotheses: (i) children can reason successfully by mathematical induction, and (ii) such reasoning is modal, i.e. acknowledged to be necessary;
- its empirical basis is in a recent study of numerical reasoning by children ($n = 100$) aged 5–7 years in school Years 1 and 2. This study was an adaptation of Inhelder and Piaget's novel claims shown to be at variance with recent positions in children's mathematics learning. Their two hypotheses were investigated along with a third hypothesis about (iii) children's knowing when-to-count and when-to-reason. The evidence was analysed in three ways. One was a non-parametric analysis of children's responses. A second was an epistemological analysis of children's reasons for these responses. A third was a logistic re-analysis of the responses using SABRE (statistical analysis of binary recurrent events) with a view to an independent confirmation. This evidence was interpreted as providing support which was strong for (i), partial for (ii), and reasonable for (iii);
- its conclusion is a reconciling position which combines the causal context of action and the normative control of reasoning in a unified epistemological framework. Its educational relevance to reasoning in the early years mathematics curriculum was discussed through a constructivist perspective about the individualization of inter-subjective knowledge.

In short, the argument contributes to filling an important gap in current research on mathematical development during childhood.

Chapter 1

Introduction

Mathematical induction is a standard form of argument in mathematics and is widely used to advance mathematical knowledge. A case in point was Andrew Wiles' demonstration of Fermat's last theorem. Providing a proof of that theorem was a major advance since it was an outstanding problem which had defied mathematicians up to 1994. Wiles' achievement was to supply a proof. And his proof was based on reasoning by mathematical induction. His starting point was the Taniyama–Shimura conjecture about the relationship between elliptic equations and modular forms.

> The challenge for Wiles was to construct an inductive argument which showed that each of the infinity of elliptic equations could be matched to each of the many modular forms. Somehow he had to break the proof down into an infinite number of individual cases and then prove the first case. Next, he had to demonstrate that, having proved the first case, all the others would topple.
>
> (Singh, 1997, p. 232)

Wiles' proof relied on the domino effect of induction in mathematics whereby if something is demonstrated to be true of one case and its successor, it is true of all cases in the series.

The central question in this study is how reasoning by mathematical induction develops in the human mind. Adult mathematicians are capable of reasoning by mathematical induction. Is this an accomplishment of childhood or adolescence? And either way, how does the development of reasoning by mathematical induction take place?

The present study has its starting point in the dispute at the turn of the last century between Gottlob Frege and Bertrand Russell, on the one hand, and Henri Poincaré on the other. Frege and Russell argued that reasoning by mathematical induction is entirely logical. Poincaré denied this, maintaining instead that mathematical intuition is required, and not merely logic, in inductive reasoning in mathematics. If Poincaré was right, reasoning by mathematical induction is neither inductive reasoning in science nor deductive reasoning in logic. But if Frege and Russell were right, reasoning by mathematical induction would be manifest with the onset of logical reasoning. There is an issue of substance here with direct implications for developmental psychology and education.

The discussion ahead is as follows. The central part of the present study deals with a recent empirical project concerned with reasoning by mathematical induction on the part of children aged 5–7 years. This empirical project is reviewed in Chapter 6. Anyone with empirical interests may well be inclined to jump straight to this section. But this would be to miss the principal question: which is the better position — Poincaré's or Russell's — about reasoning by induction in children's mathematics? This question is contextualized in Chapters 2–5. The review in Chapter 2 is a critical analysis of two answers to the question "What is reasoning by mathematical induction?" Its main conclusion is that a developmental epistemology is required, which opens the door for Piaget's research programme which is reviewed in Chapter 3. Central to his case was a study of reasoning by mathematical induction by children aged 5–8 years due to Inhelder and Piaget. Their study presented evidence to show that children can reason by mathematical induction on the basis of iterated action and that their reasoning is modal. In Chapter 4, a review of research in developmental psychology relevant to children's inductive and deductive reasoning shows that their study has been neglected even though alternatives are not to hand. A methodological argument is presented in Chapter 5 for the joint collection of evidence about children's responses and their reasons for these responses. In Chapter 6, the design of the present study based on a sample ($n = 100$) of children aged 5–7 years is reviewed in relation to three hypotheses with school Year as the independent variable concerning: (i) children's reasoning by mathematical induction based on iterated action; (ii) children's modal reasoning based on logical necessity; (iii) children knowing how to count without knowing when to count. The findings are reviewed in terms of an interdependent analysis based on both quantitative data dealing with children's responses and qualitative data dealing with their reasons in five categories for these responses. In Chapter 7, an interpretation is proposed, starting from seven main conclusions which are discussed through the three hypotheses. These conclusions are taken to provide direct support for Inhelder and Piaget's conclusion that children aged 5–7 years can reason by mathematical induction; indirect support for their conclusion that these children understand the necessity of this inference; and general support for the independence of children's knowing how and when to count. These conclusions lead to an epistemological framework in which the unit of analysis is an act of judgment. This unit of analysis combines "causal facts" and "normative facts". As such, it reconciles the psychological investigation of the causal conditions of thought and the epistemological analysis of the reasons which are its grounds. A possible mechanism is proposed in terms of the interplay between an agent's successive actions of response-making and reason-giving. In Chapter 8, a general overview of Piaget's educational model is reviewed along with six specific implications arising from the present study. There is an opportunity, which is currently missed, to promote reasoning by mathematical induction — indeed, reasoning in general — throughout the whole of schooling. Finally, a commentary (due to Damon Berridge) is provided in Chapter 9 which turns out to provide general confirmation of the main conclusions based on an independent re-analysis using SABRE (statistical analysis of binary recurrent events) of the quantitative evidence available in this study.

Chapter 2

Mathematical Induction

Mathematical induction has two interesting properties. First, it involves a necessary inference. Second, this inference is a generalization from particular to general. This means that mathematical induction has paradoxical properties: it is similar to both logical deduction and empirical induction and yet is different from both.

2.1 Mathematical Induction, Logical Deduction, and Empirical Induction

There are several recent surveys of research on human thinking and reasoning through logical deduction or empirical induction (Falmagne & Gonsalves, 1995; Garnham & Oakhill, 1994; Holyoak & Spellman, 1993; Johnson-Laird, 1999). At issue here is how these two standard forms of reasoning are different from mathematical induction, and so some examples may be helpful (see Box 1.1).

The first example in Box 1.1 concerns mathematical induction. The premises of this argument are shown above the line and the conclusion is shown below it. All reasoning requires some inference. What differs is the nature of this inference. In this example, the premises are particular equalities which are true in these three cases where the sum of the numbers on the left-hand side is equal to the squared number on the right-hand side. The inference in this example has two properties since it is a *necessary* inference from these premises to a conclusion which is a *generalization* true of all integers (Douglis, 1970). The validity and power of reasoning by induction is widely accepted in mathematics (Borowski & Borwein, 1989; Daintith & Nelson, 1989). This example can now be used to bring out why the two properties (necessitation, generalization) make mathematical induction distinctive.

The second example in Box 1.1 is deduction. A valid deduction in logic is truth-preserving, i.e. if its premises are true, then its conclusion must be true as well (Quine, 1972, p. 4). The same point was clearly expressed by Haack (1978, p. 22) in that logically valid reasoning is such that it couldn't have, not just doesn't have, true premises and false conclusion. Quite simply, valid arguments are necessarily truth-preserving (Sainsbury, 1991, p. 15). The conclusion (*there is red wine on the table*) is not itself a necessary proposition. The reason is simple. It is not necessary for there to be red wine on the table. Even so, this conclusion is necessitated by these premises. This is because on condition that these premises are true, this conclusion must be true (Quine); this conclusion could not be otherwise (Haack); the truth of this conclusion

---

Box 1.1: Examples of reasoning by mathematical induction, logical deduction, and empirical induction

Mathematical induction

Premises

$1 = 1^2$
$1 + 3 = 2^2$
$1 + 3 + 5 = 3^2$

therefore

Conclusion

$1 + 3 + \ldots + (2n - 1) = n^2$

Logical deduction

Premises

there is either red wine or white wine on the table
there is no white wine on the table

therefore

Conclusion

there is red wine on the table

Empirical induction

Premises

this raven is black
that raven is black
all observed ravens are black

therefore

Conclusion

all ravens are black

---

is necessitated by the truth of these premises (Sainsbury). In other words, valid deduction is necessitating.

The third example in Box 1.1 concerns empirical induction, where the inference is a generalization about a property true in all cases in a specified class on the basis of instances of that property true of some cases in that class. This fits the narrow (generalizing) sense of induction identified by Skyrms (1995) which he contrasts with the broad sense of an ampliative inference, i.e. the inference makes claims which go beyond the claim jointly made by the premises. Examples of induction in this broad sense include argument by analogy and inference to causes from signs. In this example, the fact that there are instances of a properties (black) which are true of a class (raven) is a past association which does not guarantee its continuation — witness the discovery of black swans by Europeans whose prior experience had been limited to white swans. It is always possible that the regularity may fail (Russell, 1912/1959), and that a predicate successfully applied may fail in the future (Goodman, 1979). This means that even though empirical induction is generalizing, it is not necessitating. Repeated co-instantiation is not the same as inferential necessity

And this is why mathematical induction is a special case. It is similar to deduction in that an inference is necessitating. It is similar to empirical induction in that an inference is generalizing. Using Waismann's (1951, p. 92) example, it is easy to show that

$$1/3 = 0.3$$

and that

$$1/3 = 0.33$$

and that

$$1/3 = 0.333$$

It would be tedious to repeat this calculation. Yet if this is not done, how is it known whether some digit other than three will turn up in an expansion to 10, 100, or 1000 decimal places? Indeed, it would be impossible to do this calculation for an infinite number of cases. Yet this is what

$$1/3 = 0.333 \ldots$$

means, namely that one and the same number recurs such that this number is three. Such an inference is due to mathematical induction. And this means that the inference is both something novel, which is over and above what is said in the premises, and something necessitated by its premises.

2.2 Poincaré's Analysis of Mathematical Induction

A definition of mathematical induction in arithmetic was stated by Poincaré which he characterized as reasoning by recurrence:

> if a property is true of the number 1, and if it is established that it is true of $n+1$ provided it is true of n, it will be true of all whole numbers. (Poincaré, 1908/1952, p. 149).

This definition includes two conditions. One is a base property about a specified property true of a particular number. The other is a recursive property such that if the base property is true in one case, it is also true of that number's successor. It is not sufficient merely to assert that these properties are true in particular cases. Rather, they must be proved to be true in those cases, i.e. their truth is demonstrated in a theorem (Waismann, 1951, p. 91; cf. Singh, 1997, p. 232). A mathematical induction in arithmetic is an inference from the base property together with a recursive property demonstrated to be true of particular integers to a generalization true of all integers.

This definition of mathematical induction secures the two distinctive properties of necessitation and generalization. The definition secures a generalization (universalization). The reference in the premises is to a property true of *any* (some particular) number. But the reference in the conclusion is to *any* number whatsoever, any number at all. Note that the term *any* is ambiguous here since it covers an inference from some-to-all. But there is a crucial difference as well which is the second point. Empirical induction is empirical, and so is not logically necessary. Any empirical generalization (all the butter that I eat is English butter) as to any natural necessity (all butter melts at 150°F) could have been otherwise. But this means that neither is a logical necessity since, by definition, a logical necessity could not be otherwise (Smith, 1997). A true (empirical) generalization could have been otherwise — my butter could have been Danish. A natural necessity could not be otherwise in view of the way the actual world is, but would have been otherwise if the world had actually been different. By contrast, a logical necessity could not be otherwise in any world at all (Haack, 1978; Sainsbury, 1991). Poincaré's definition states a necessitating — not an empirical — property of numbers, i.e. any number to which both conditions are applicable must have the specified property. Evidently mathematical induction combines the two conditions to yield valid inferences about the necessarily true properties of mathematical objects. It is this combination which ensures that mathematical induction is a potent form of reasoning.

But what is mathematical induction? What exactly does mathematical induction amount to? This question is especially pointed in view of the joint possession in mathematical deduction of two properties which are exclusive properties of logical deduction (necessitation) and empirical induction (generalization). Further, how is mathematical induction understood by human minds? Any such induction requires an inference true of an infinite number of cases by a human mind confined to a finite number of cases. There are major problems about knowledge and reality here. As such, they are central to philosophy which includes both ontology and epistemology.

One part of the problem is ontological and concerns what reality is like. This is the problem of "what there is". Ontological problems are present in all domains, for example in sociological disputes as to whether society is an aggregation (Comte) of individuals or whether society is something over and above (Durkheim) the individuals composing it (Piaget, 1977/1995). The ontological problem is whether or not there are such things as societies. The other problem is epistemological and concerns whether and how there can be human knowledge of reality at all. Pareto (1963, para. 972) pointed out that "at bottom what people want is to think — it matters little whether the thinking be sound or fallacious". The problem is how any human mind can gain an objective understanding of the world when the human mind has a propensity for wishful thinking which masquerades as objectivity. Ontological and epistemological problems arise in mathematics. Thus the question "Do numbers exist?" is an ontological question. The point about this question is that any affirmative answer leads straight to the question "How is it possible for numbers to exist since they are plainly not part of the actual world?" Frege (1884/1950, pp. 41, 49) pointedly argued that a number, such as the number one, is unique and so could not be identified with any of the indefinitely many particular objects in the actual world. The actual world includes indefinitely many objects, such as the Eiffel Tower or the single slice of bread on a plate and many others. So how can each of these be the unique number which is one? The question "How can numbers be known by human minds?" is an epistemological question. Frege (1884/1950, pp. 28, 59) argued that any actual object could not be the source of human knowledge of numbers since the possession of numbers is required for anyone to make numerical statements about them in the first place. This is because any actual thing can be both "one and many". For example, hearing The Beatles is an insufficient basis for making numerical statements, such as "This is one group" or "This is four musicians, John, Paul, George and Ringo". The same experience is compatible with a statement about *one* group and *four* musicians. Yet the number one and the number four are completely different numbers. Thus these ontological and epistemological problems in the philosophy of mathematics are fundamental. However, they are jointly intractable since there is no agreed answer to both problems (Benacerraf, 1973; Katz, 1995).

Poincaré's argument was as follows. Mathematical reasoning is creative and so is generative of new knowledge. But logic is sterile and cannot be generative of new knowledge. Logic provides verification which "differs from proof precisely because it is analytical and leads to nothing [whereas] mathematical reasoning has of itself a kind of creative virtue and is therefore to be distinguished from the syllogism" (Poincaré, 1902/1905, pp. 3–4). It follows that mathematical induction is not reducible to logic. His proposal was that its basis lies in mathematical intuition in conjunction with logic. His view was that mathematical induction is an "instrument of transformation [which] contains, condensed, so to speak in a single formula, an infinite number of syllogisms" (Poincaré, 1902/1905, p. 9). He also made an explicit commitment to constructivism in that mathematicians are reckoned by Poincaré (pp. 15–16) to "proceed 'by construction' although a construction only becomes interesting when it can be placed side by side with other analogous constructions". A key distinction was also invoked between instructive and productive construction in that (mathematical discovery) "does not consist in making new combinations with mathematical entities that are already known [but rather] in not constructing useless combinations, and so in constructing those that are useful" (Poincaré, 1908/1952, pp. 50–51). Central to instructive construction is mathematical intuition which Poincaré (p. 55) regarded as "sudden illumination". In this regard, Poincaré is in good company. Comparable references to the importance of intuition in mathematical reasoning were made by his predecessors, such as Kant (1933), and his successors, such as Gödel (Wang, 1996). But this is to invoke authority. The real point is whether Poincaré's argument is sound.

Poincaré's analysis of mathematical induction, or reasoning by recurrence, made three points. One is that numbers have recurrent properties, which are responsible for their domino effect (Singh, 1997). Frege (1879/1972) and Russell (1919) had characterized these as the hereditary properties of numbers, so there is common ground here. Second, Poincaré (1908/1952, p. 48) pointed out that "it is not in the passage from premises to conclusion that we are in danger of going astray. But between the moment when we meet a proposition for the first time as the conclusion of one syllogism, and the moment when we find it once more as the premises of another syllogism". His point was that drawing a conclusion from a set of assembled premises is not the same as assembling a chain of premises–conclusions in a complex argument in the first place. This is a good point. Although Couturat (Mooij, 1966, p. 71) had contended that the mathematical inference is a logical deduction, this misses the point as to how specific premises are identified for the passage from one condition (the base property) to the other (its recursion) as a precondition of making an generalizing and necessary inference. Russell (1919, p. 6) had conceded this point, at least with respect to Peano's fifth axiom which blatantly amounted to "the principle of mathematical induction". This is a vicious circularity. And this is where mathematical intuition has a place, in Poincaré's view. So much is evident in his contention that mathematical intuition is "the perception of a whole argument at a glance" (Poincaré, 1908/1952, p. 49). It is this intuition which secures the passage from the conclusion of one argument (*syllogism* is Poincaré's term) as it becomes the premise in another. This is a constructive shift. And this phenomenon is not in doubt, even if there is some debate as to its nature. This "passage" is not analytic, still less is it a logical deduction. According to Burge (1993), it is better regarded as preservative memory, which has escaped the attention of cognitive scientists. At any event, Poincaré (1908/1952, p. 128) drew the conclusion that "the principal aim of mathematical education is to develop certain faculties of the mind, and among these intuition is not the least precious". Third, Poincaré (1902/1908, p. 15) contended that "mathematicians therefore proceed 'by construction', they 'construct' more complicated combinations". The scare-quotes are due to Poincaré who was not impressed by the metaphor of the building site (cf. Chandler, 1997; Scholnick, 1999; Smith, 1999e). But his use differs from that of another French logician who later argued that a logical deduction is the linking of a consequence to a hypothesis in as much as "the consequence is constructed with the hypothesis. Constructive operations are not operations in the mind, but are operations carried out mentally" (Goblot, 1929, sect. 169). One way to interpret this contention is in terms of the distinction between entailment and inference. The entailment relation linking premise and conclusion in a logical deduction is necessary (Marcus, 1993; Smith, 1999d). What links a belief in these premises and a belief in this conclusion is an inference. And this inference is typically not necessary, or at least not in young children's reasoning (Vygotsky, 1986), and maybe not even in adult reasoning (Stroud, 1979). According to Goblot, this link is a construction. Poincaré exploited this distinction to good effect. Even if a premise logically entails a conclusion, a human inference of this conclusion from this premise need not itself be logical. There was a further argument. Not only is construction not bound to be logical, the "new" logic itself presupposes number, argued Poincaré (1908/1952, p. 164), in that propositions have "two" truth-values (truth, falsehood), whilst the null class has "zero" members. Far from mathematics being reducible to logic, the suggestion is that mathematics is required by logic in the first place.

Even so, there are difficulties in Poincaré's account. Couturat observed that although Poincaré was aware of the "new" logic, he was not adept in its systematic use (Mooij, 1966, p. 71). Recall his reference to logical syllogisms which were merely a special case of a more general logic due to Frege and Russell. According to Kenny (1995, p. 209), "if Aristotle was the founder of logic, Frege refounded it, and logic has developed faster and further in the period between his time and the present day than it did in all the centuries which separated him from Aristotle". And Hintikka (1996, p. vii) has remarked that Russell's (1903/1964) independent work was a major step forward in the liberation of traditional approaches to logic and the foundations of mathematics. This means that Couturat's observation on Poincaré's relative lack of expertise in the "new" logic is interesting since the implication is that Poincaré's analysis is liable to be deficient in missing the logical point. Indeed, Poincaré (1902/1905, p. 2) had traditional logic in mind in his claim about its sterility. He took logic to be analytic, amounting to a complicated way of stating that A is A. On this view, logic is a tautology. Since mathematics is creative, it must differ from logic. According to Poincaré, this meant that mathematics is synthetic *a priori*. This is effectively an acceptance by Poincaré of Kant's position that reasoning in mathematics is based on principles which are both synthetic (they fit reality) and *a priori* (they are necessary). But this view had already been criticized by Frege (1884/1950, p. 6) who had rejected Kant's position in favour of the position that these principles are analytic. Frege's point was that not all analytic truths are self-evident and obvious — as he supposed Kant supposed — and so productive reasoning in mathematics can result in novel knowledge. Crucially, Poincaré (1908/1952, p. 125) had made the arresting claim that "logic breeds monsters". By *monsters*, Poincaré meant contradictions. Indeed, Poincaré (p. 194) took some delight in noting that the "new" logic is not barren due to its production of contradictions such as Russell's class paradox (1919; see Smith, 1993, p. 29). The implication is either that logic is monster-prone and so is an inadequate foundation for mathematics, or that mathematics should have a more secure foundation elsewhere. Both alternatives are bleak. Lakatos (1976) has shown that mathematical reasoning is equally monster-prone, and so in this regard is no different from logic. This effectively neutralizes the first alternative. The second alternative is vacuous unless its specific character is elaborated. This is something which Poincaré did not carry through. His interpretation of intuition was in terms of "sudden illumination". But this is a visual metaphor, not a veridical explanation (Kitcher, 1983, pp. 50–51). Further, it does too little to clarify the key distinction between a mathematical generalization and an existence statement in mathematics. The negation of a generalization is not the same as the statement of a correct counterpart, and so the outstanding question is how "mathematical intuition" secures the advance from the first to the second (Waismann, 1951, pp. 97–98). Indeed, this objection is formulated with some cogency by Frege whose own position is now reviewed.

2.3 Frege's and Russell's Analysis of Mathematical Induction

Frege and Russell gave a reductionist answer according to which mathematical induction is — is reducible to — deductive reasoning in logic. Their answer was a version of realism. Poincaré had denied this in his version of constructivism. These are contrary interpretations and so both cannot be right. According to Dummett (1981, p. xxxviii), constructivism continues to be the standard rival to realism. It is common ground that mathematical knowledge does develop and that reasoning by induction was a principal form of reasoning in mathematics (Frege, 1897/1979, p. 165; Poincaré,

1908/1952, p. 149; Russell, 1919, p. 27). But this agreement does nothing to settle the philosophical dispute about what mathematical induction really is and how it can be put to sound use.

Modern logic is due to Frege. The scale of his contribution has been well expressed by Kenny (1995) and was quoted in section 2.2. Central to Frege's advance were two principles. First, he defined logic as the formal science of truth (Frege, 1897/1979, p. 128). Second, he saw in logic the means "to break the power of the word over the human mind" (Frege, 1879/1972, p. 106). Frege put forward his system of logic as a unitary and complete way in which all truth-preserving inferences could be inventoried. Frege's logic was extensional (Haack, 1978; Quine, 1972; Sainsbury, 1991). Further, Frege believed that this logic fits all mathematical inferences (Dummett, 1991). And Frege's position has special relevance to the nature of mathematical induction in three respects.

First, Frege argued that traditional logic from Aristotle's *Prior Analytics* (1987) to Boole (1854/1958) was incomplete. There is a failure to distinguish subordination and subsumption. The credit for first drawing this distinction in logic is due to Frege, stated Russell (1903/1964). Traditional logic did not extend to mathematical propositions. The signs *0* and *1* are used in Boolean algebra without a mathematical meaning, and are not used at all in syllogistic logic. Frege's logic was designed to express mathematical propositions. This included one important property of mathematical truths, their generality expressible through logical quantification. Frege (1881/1979) regarded this as a non-negligible benefit. Others have regarded this as his highest achievement (Kneale & Kneale, 1962).

Second, Frege (1884/1950, p. 19) expressed scepticism about intuition in mathematics. His point was that the sense of this term is left either ambiguous or opaque, notably in Kant's account. And Poincaré had sided with Kant here. Is an intuition of a pack of cards an intuition about one pack or about 52 cards (Frege, 1884/1950, p. 28)? There is nothing in the content of an intuition to favour one of these rather than the other. One-and-the same object can serve in an intuition about one pack or about 52 cards. Thus number is presupposed in forming either of these intuitions and so intuition cannot be the basis of number.

Third, Frege had formulated an alternative interpretation of number, namely logicism in which any putative concept in mathematics can be given a completely logical analysis (Frege, 1884/1950, p. 80). Frege believed that this requirement is met in his definition which utilizes the relation of one–one correlation and the extensional equivalence of logical classes. His waiter example is instructive:

> If a waiter wishes to be certain of laying exactly as many knives on a table as plates, he has no need to count either of them: all he has to do is to lay immediately to the right of every plate a knife . . . Plates and knives are thus correlated one to one, and by the identical spatial relationship. (Frege, 1884/1950, p. 81)

It is possible to know that there are *as many* knives as forks without knowing *how many* are on the table. Frege concluded that mathematical induction really amounted to deductive reasoning in logic. "The present work will show that even inferences which on the face of it are peculiar to mathematics, such as that from n to $(n + 1)$, are based on the general laws of logic, and that there is no need of special laws for combinative thought" (Frege, 1884/1950, pp. iv, 118). Under this account, the inference from n to $(n + 1)$ in mathematics is really a deductive inference in logic.

Russell (1903/1964, pp. 5, 113) had a comparable view in each of these respects. In so doing, he exploited a brilliant example. The months of the year, Christ's apostles and Napoleon's marshals can be placed in one–one correspondence, even though the members of these classes have nothing else in common. October is not a marshal, Mark was not Marshal Ney. Since the members of these classes can be placed in a one–one correlation, it is therefore possible to establish that there are *as many* members in each class without knowing *how many* members there are. More generally, "mathematically, a number is nothing but a class of similar classes" (Russell, 1903/1964, p. 116; 1919, p. 18). On this account, 12 is the class of all the classes of dozens. This account is non-circular since the concept of number, which is mathematical, is not used in its own definition, which is logical. In short, the logicist proposal is that any numerical proposition can be recast as a non numerical proposition This means that, in principle, this reduction can always be carried through, even if numbers are in practice not reduced to logic.

2.4 Critical Review

In the sequel, the Frege–Russell position became the dominant position. Even so, the adequacy of this position has not gone without challenge. Eight objections are now briefly reviewed.

First, logicism requires a position about the status of definitions. Russell denied that definitions have a truth value. They are volitions, not proposition (Whitehead & Russell, 1910/1970, p. 11). Yet Frege (1892/1960) did assign truth values to identities of which definitions are a paradigm case. Thus logicism is not a unified position and an ambiguity is evident with adverse implications for the objectivity of mathematical knowledge.

Second, logicism is problematic because of the class paradox which was initially formulated in a letter which Russell (1919, p. 136) sent to Frege (1980) in 1902. This paradox can be explicated by analogy with a library catalogue listing catalogues which do not list themselves (Hersh, 1997). It vitiated Frege's system of mathematics with adverse implications for Russell's own system. This was the "monster" in the new logic referred to by Poincaré. Logicism is not itself a paragon of logical rectitude.

Third, Gödel's theorem shows that no formal system rich enough for elementary arithmetic can contain a joint demonstration of its own consistency and completeness (Wang, 1996). At a stroke, logicism is undermined. If the logical foundation to which mathematics is reduced by logicism is either inconsistent or incomplete, so much the worse for logicism. This does not mean that a logicist interpretation of mathematics actually is incomplete or inconsistent (Hersh, 1998; Singh, 1997). But it does mean that this possibility cannot — cannot in principle — be excluded. Thus logicism as a foundation for mathematics is inherently vulnerable to potential collapse.

Fourth, whatever numbers really are, they are not sets (classes). In Benacerraf's (1965) thought experiment, two children learn mathematics strictly along logicist lines, but they do this in different ways by learning different, though equally valid, formulae in set theory. But then the question "Which sets are numbers?" is insoluble. Both children could invoke valid formulae in set theory for incompatible answers as to which numbers these are. This means that numbers have properties which cannot be determined by the properties of sets. From this it follows that numbers cannot be sets.

Fifth, logic is not itself a unity. So the question "Which logic?" is acute. Even first-order logic has variants (Haack, 1978; Quine, 1972). And modal logics are not first-order logics anyway (Marcus, 1993; von Wright, 1983a). Crucially, it is now argued that first-order logic is "too weak for most mathematical purposes [and notably for] mathematical induction" (Hintikka, 1996, p. 7). And deontic logic is a logic of action which is devoid of truth-value altogether (von Wright, 1983a).

Sixth, there is no account which adequately answers two questions "What is mathematics?" and "How are mathematical truths known?" in one account (Benacerraf, 1973). Sponsors of realism, nominalism and constructivism are to hand (Dummett, 1991; Field, 1980; Katz, 1995). But these are three distinct perspectives. In consequence, one adequate account is still awaited.

Seventh, the Kantian principle "ought implies can" has an epistemological application (Kornblith, 1985). If an epistemological account which states how knowledge should be acquired has the consequence that knowledge could not be acquired in some population, which is independently regarded as a community of knowers, then the account is itself wrong. This argument has the consequence that a non-empirical epistemology is not sufficient. Yet logicism is tied to a non-empirical epistemology which is silent about the ontogenesis of mathematical reasoning. It was never made entirely clear in logicist accounts how mathematical knowledge is generated, or even whether human knowledge is possible at all. Russell (1959) was himself prepared to accept the sceptical consequences of his work. By contrast, Frege (1918/1977) was committed to a positive epistemology about human knowledge. Once again, there is a split in their commitment to logicism. Further, this split is once again important: the non-possession of human knowledge in mathematics is a reason for rejecting any account which has this as its consequence.

Eighth, the view that mathematical knowledge is *a priori* in character is now argued to be false (Kitcher, 1983). The implication is that such knowledge is due to empirical learning. But what does that mean? It is common ground that all mathematical truths are necessities (Smith, 1997, 1999d). It is also common ground that any necessary truth can be learned empirically (Kripke, 1980; Leibniz, 1981). But these two admissions leave open the question about the specific character of this learning. How is it possible for there to be empirical learning of a necessary truth? Mathematical truths have non-empirical properties, such as necessity and generality. It is not known how an understanding of such properties is acquired on the basis of temporal learning (Bickhard, 1988; Kitchener, 1986; McGinn, 1991; Nagel, 1997). This major problem which was well known to Plato is currently unresolved.

2.5 Conclusion

The main conclusion is clear. Logicism was one of the major programmes in mathematics. Current programmes in mathematics continue to be defined by reference to logicism (Dehaene, 1997; Hersh, 1997; Nagel, 1997; Singh, 1997). Further, the refounding of logic was one of the major successes of logicism (Hintikka, 1996; Kenny, 1995). Even so, the case for logicism remains, at best, not proven and, at worst, flawed. In particular, Poincaré's dispute with Frege and Russell remains unresolved with regard to reasoning by mathematical induction. Is there an alternative?

The answer is "yes". There are two reasons why and this classical dispute is central to both. The general reason concerns constructivsim. It was noted above (cf. Dummett, 1981) that constructivism (Poincaré) is the chief rival to its alternatives which are realism (Frege and early Russell) and nominalism (late Russell). But there are many versions of each these three (philosophical) positions. Recent versions of constructivism in psychology and education continue to attract *attention* (Larochelle *et al.*, 1998; Phillips, 1997; Steffe & Gale, 1995). A good question here is which version of constructivism commands *agreement*. What seems to be missing from current discussions is recognisance of realism and nominalism as the main alternatives to constructivism, not the local difficulty as to whether construction is individual or sociocultural. There is something of an *impasse* in view of major differences between sponsors of these different versions of constructivism (Smith, 1996b, 1999e). One way out is to *reculer pour mieux sauter*, to go back so as to make a better leap forward. The dispute about mathematical induction could be instructive in this regard. The particular reason concerns Piaget's constructivism. Piaget's account of logic and mathematics was set out expressly in relation to the positions taken by Poincaré, Frege, and Russell (Ducret, 1984; Smith, 1999a; Vidal, 1994). And it was their dispute over mathematical induction which led to the design of an important but neglected study by Inhelder and Piaget in 1963. This link will be elaborated in Chapter 3. Crucially, their study led to the conclusion that reasoning by mathematical induction develops during early childhood. This is a spectacular claim. This type of reasoning is virtually absent from current research on children's development. And this type of reasoning is a notorious trouble-spot for adolescents and adults in educational settings. The discussion in this chapter may well be interesting in its own right. But it would have added value if it can throw light on the development of reasoning by mathematical induction during childhood.

Chapter 3

Reasoning by Mathematical Induction: Piaget's Critique

Piaget's research programme was distinctive in four respects. First, Piaget (1950, p. 36) invoked the "ought implies can" principle well in advance of Kornblith (1985). Second, Piaget (1950, p. 8) also anticipated Kitcher's (1983) general argument when he pointed out that the demarcation of *a priori* from *a posteriori* questions could not be carried through on a principled basis. In that case, the question would remain open as to whether an epistemic instrument is "actually at the subject's disposal. Here, whether we like it or not, is a question of fact" (Piaget, 1970/1977, amended translation in Smith, 1993, p. 7). Their combination led to a third point. Kant (1933) had asked the question "How is knowledge possible?" Piaget (1950, p. 12) contended that this normative question should be replaced by an empirical question "How does knowledge develop?" Although a developmental epistemology is also empirical, the converse is not the case in as much as empiricist and evolutionary epistemologies are also empirical but not thereby developmental. Taking the latter seriously leads to a fourth point, namely the extent to which knowledge is developed rationally. A developmental epistemology would focus on "the development (*genèse*) of rational operations in the child" (Piaget, 1941, p. 216). Rationality is a normative notion. What counts as sound or fallacious reasoning is determined by logical norms, even if all people in fact make inferences on the basis of pseudo-logic (Pareto, 1963). A developmental epistemology could contribute empirically based answers to these normative questions based on the study of "normative facts", i.e. by reference to the progressive use of norms through time, whether in individual minds or through cultures (Piaget, 1977/1995). In short, Piaget's research programme required a developmental and empirical epistemology directed on "normative facts" (Smith, 2002a,b). It was this developmental epistemology which was applied by Piaget to reasoning by mathematical induction. This is because although logicism was an influential position, Piaget believed that Poincaré had made a good point.

3.1 Piaget on the Analysis of Frege and Russell

Piaget was introduced to Frege's work as a student in Neuchâtel (Smith, 1999a) and was soon to make explicit use of Russell's work (Piaget, 1923). There were parallels between their positions and his. But there were objections by Piaget to their positions as well.

One parallel concerns equality which is the central relation in arithmetic.

> If you drop equality from arithmetic, there's almost nothing left. (Frege, 1903/1979, p. 165)

This is an important remark. Equality is identity, and this means that a (numerical) equality and a (numerical) identity are one and the same thing (Frege, 1884/1950, p. ii; 1892/1960, p. 56). In turn, this means that the relation of identity cannot link two (or more) objects since two objects are thereby not one-and-the-same object. This is why Frege interpreted identity as a relation linking (two or more) linguistic expressions. The sense of these expressions may differ, even though they have (one and) the same reference. This is epistemologically important because true identities have different cognitive values (Frege, 1892/1960, p. 78). An identity statement may express a tautology amounting to ($a = a$):

the Morning Star is the Morning Star.

But the cognitive value of other identity statements is different, amounting to ($a = b$):

the Morning Star is the Evening Star.

Perplexed by the latter, someone might say "But how can two things be one?" Armed with Frege's distinction, the reply would be "But there is only one thing here, the planet Venus which is the unitary reference of the two expressions *Morning Star* and *Evening Star.*" The point is that there can be different modes of epistemological access to — different ways of getting to know — a mathematical equality.

Piaget was familiar with this line of argument (Smith, 1999a; Smith *et al.*, 1997). And it was central to his account of number. Piaget and Szeminska (1941/1952, p. 238) asked whether "children are capable of understanding the identity (*identité*) of a whole through different additive compositions of its parts $(4 + 4) = (1 + 7) = (2 + 6) = (3 + 5)$". In their study, this identity was contextualized in terms of a child who was given a snack which was divided on day I equally into a morning snack (4) and an afternoon snack (4), but divided unequally on day II into a morning snack (1) and an afternoon snack (7). Just as in Frege's example an astronomer in one cultural context may or may not believe that the Morning Star is identical to the Evening Star, so too in this psychological context a child may or may not believe that the snacks on day I equal the snacks on day II. It is for this reason that Piaget and Szeminska (1941/1952, p. 244) interpreted children's development as a progression from "the inequality (*inégalité*) of the parts (snacks) to their equality (*égalité*)", and so that "identity (*identité*) is the outcome not the source of reversibility" (p. 259). Thus Frege reckoned that equality is the constitutive relation in arithmetic, whilst Piaget and Szeminska assigned it comparable importance in their study of the ontogenesis of number.

A second parallel is the denial that arithmetic has its basis in counting. In the waiter example quoted in Chapter 2, Frege had pointed out that it was not necessary for the waiter to count *how many* knives and plates are on the table to see whether the number is the same. It is sufficient to place a knife on the table next to each plate, enough to show that there are *as many* knives as plates. The implication is that logical reasoning, and not counting, can be the basis of numerical knowledge and so counting is not necessary for an understanding of number. Further, Russell pointed out that counting is not sufficient either. The natural numbers (1, 2, 3. . .*n*) are used in counting, and so cannot be defined through counting without circularity (Russell, 1919, p. 15). So this is a second reason why counting is not the basis of number.

Although some developmentalists (e.g. Gelman, 1978; Gelman *et al.*, 1986) have deplored what they believe to be the downgrading of counting knowledge, for example by Piaget, they have missed the main point that mathematical reasoning about "as many" (equality) is at least as important as, and is arguably more important than, knowing "how many". Yet exactly this point was central to the logicist account of Frege and Russell. Children may well learn to make such statements as

$$1 + 1 = 2$$

$$1 + 2 = 3$$

$$1 + 3 = 4$$

before they go to school. Yet numerical understanding requires both the sense and reference of the expressions in such statements to be fixed. Reverting back to the snacks task, children who reason that there is more on day II than day I are effectively denying the inequality

$$4 + 4 = 1 + 7$$

This error is compounded when children (see Dini's protocol: Piaget & Szeminska, 1941/1952, p. 241) justify this inequality on the grounds that, over day I and II, the Morning Snack is $1 + 4 = 5$ whilst the Afternoon Snack is $7 + 4 = 11$. The implication is that since

$$5 \neq 11$$

therefore

$$4 + 4 \neq 1 + 7$$

One interpretation of this evidence is that the reference of these expressions is not fixed in the minds of these children. It is for this reason that their knowledge of counting is an inadequate basis of numerical reasoning.

Third, Frege regarded language as a major source of conceptual confusion which could be dispelled only by logical analysis. The numeral *one* is not the name of an object just because each object is an object, and this multiplicity is incompatible with the unity of one (Frege, 1884/1950, p. 41). According to Frege (1879/1972, p. 106), logic acts as a corrective to language in human thought. Russell (1905/1992) had a similar view in that paradigm of philosophy which is his theory of descriptions. It is no doubt for this reason that logic was central to Piaget's model of cognitive development. Children are able to use language from the earliest years of childhood. Much less clear is whether they have a comparable mastery of logic. It is this logicist position which was exploited by Inhelder and Piaget in the design of their study by characterising mathematical intuition in terms the human capacity for iterative action. It is their proposal which is now reviewed in section 3.2.

Even so, Piaget (1942) rejected logicism, notably in his critique of Russell's account of number. This critique is theoretical and sets out a logical model of cognitive development based on group theory. As such, this model is well defined. Notice the date of this critique which has not been translated into English and which is simply presupposed in a concurrent, empirical critique (Piaget & Szeminska, 1941/1952). Notice also that the psychological evaluation of this latter has tended to focus on Piaget's *tasks* rather than on his *model* (Bideaud, 1992; Case, 1999; Nunes & Bryant, 1996; Siegler, 1996). This is to miss Piaget's point that mathematical reasoning is *not* reducible to logic.

Piaget's critique was primarily directed against Russell's account. His main objection was that Russell's account has pessimistic implications for children's understanding of number because children do not in fact have the expertise of a sophisticated logician. This objection is in three parts.

His first objection was that Russell's account of one–one correspondence conflated two distinct forms of correspondence, one qualitative and the other generalized. According to Piaget (1942), children can undertake qualitative correspondences, but not generalized correspondences. Yet Russell's account required the latter, and not merely the former. As an example, consider two committees (*A, B*), each of which has a chair, treasurer, secretary, and so on. These committee members can be paired in two different ways. Qualitative correspondence takes place when the chair of *A* is paired with the chair of *B*, the treasurer of *A* with the treasurer of *B*, and so on. Such correspondence is qualitative because each individual is identifiable on the basis of an antecedently identifiable property shared with one other individual. The correspondence can be similarly carried through for the remaining members of the committees. Quite different is generalized correspondence, whereby *any one term* (*n'importe quel terme*) can be paired with *any other* term at all (*quelconque*). For example, *A*'s secretary can be paired with *B*'s treasurer (Piaget, 1942, p. 90). Notice that this form of correspondence is required by mathematical induction in the logicist account. Reverting to Russell's famous example clinches Piaget's (1942, p. 235) point. The example sets out to show that the class of Napoleon's marshals and the class of months of the year have one and only one property in common in that they both have a dozen members. In other respects, marshals and months have nothing in common. Generalized correspondence requires reference to bare individuals in abstraction from their antecedent properties which have been stripped away. Thus generalized correspondence requires the members of these classes to be intersubstitutable without being identical (Piaget, 1942, p. 236). Piaget's objection is that Russell has not verified that children can in fact undertake this abstraction to stripped-down individuals. This objection is supported by the evidence due to Piaget and Szeminska (1941/1952, Ch. 3) whose study of "correspondence" (*sic*) demonstrates quite precisely that children's non-conserving responses are made on the basis of qualitative correspondence in terms of the spatial lengths of the lines. These children are not making a generalized correspondence required by Russell's account.

The second objection was that generalized correspondence is systematic in that each term is paired once; no term is paired more than once; all terms are paired; and pairing is in any possible combination (Piaget, 1942, p. 236). But systematic correspondence can be carried through only on the basis of a joint use of the logic of classes and relations. Suppose A's chair is paired with B's treasurer, i.e. a member of class A is paired with a member of class B. So this first step is based on class membership. Quite different is the next step. These two individuals should now be excluded for the next classification to be made in that some *other* member of A would next be paired with some *other* member of B. Thus the selection of the next pair is serial. But this is only possible if the individuals in each class are ordered in some way. Otherwise one individual may be paired twice over, or another not paired at all. Since this correspondence is generalized, these individuals are devoid of antecedent properties in that *ex hypothesi* the sole property in common is number, and nothing else. Thus the only property to secure their serial identification is the order of selection in each of the two classes. But ordering requires the logic of relations. It follows that generalized correspondence requires the joint use of the logic of classes and relations. According to Piaget's (1942) model, these are distinct logical operations. According to Piaget and Szeminska's (1941/1952) evidence, these logical operations are used separately — but not together — during childhood. Therefore, they are not available during childhood for the joint use which is required by Russell's account.

The third objection was that Russell's account is silent about the development of a novel ability required by his logicist account (Piaget, 1942, p. 243). This objection hinges on the distinction between logical and numerical addition (quantification), which are not the same thing. Because they are not identical, something else is required to convert the former into the latter. In Russell's account, no explanation of the basis of this conversion factor is given since the new ability is assumed to arise *ex nihil*. The starting point of this objection is clear enough. Consider two classes such as the class of skeletons of fish (Sf) and the class of skeletons of amphibians (Sa). Logical addition generates a new class which is the class of skeletons of vertebrates (Sv). Each original class (Sf, Sa) is severally a member of the latter (Sv). And so too is the joint class composed by logical addition ($Sf + Sa$). With respect to skeletons, all fish and all amphibeans are vertebrates. But numerical addition is different. Consider the numerical addition

$$12 + 12 = 24$$

The joint class which is composed by the addition $(12 + 12)$ is a member of 24. But 12 is not itself a member of 24, even though 12 is included in 24. Class membership is not class inclusion. Ironically, it was Russell (1903/1964, p. 19) who drew attention to the difference between membership, which is not transitive, and class inclusion, which is transitive, in his commendation of Frege's contribution to logic. Piaget's conclusion was that logical addition can be converted into numerical addition only on the basis of some further factor. Yet in Russell's account, this conversion occurs *ex nihil*, from nothing, amounting to a *deus ex machina* just because it has not been specified. This means that the conversion arises from nowhere unless it is specified. What is required is an account of number which fits the facts of the development of numerical reasoning (Piaget, 1942, p. 242). Piaget's proposal is encapsulated in the slogan "there will be no immaculate conceptions here" (cf. Siegler, 1996). Indeed, it is exactly this slogan which is a constitutive principle in his constructivism.

In short, Piaget accepted the importance of logic in mathematics. But he declined to accept logicism in mathematics. This is an important distinction which led to his interest in alternatives, such as Poincaré's constructivism.

3.2 Piaget on Poincaré's Analysis

Piaget (1949, p. 18) was influenced by Poincaré's work in several ways, but his own position was importantly different with regard to children's reasoning by mathematical induction.

One influence concerned the importance of reasoning by induction in mathematics. Piaget (1949, p. 382; 1975/1985, p. 67) was especially attracted to Poincaré's joint focus on rigour and creativity in the search for objectivity and novelty. Even so, he denied that logic is sterile. "Logical structures constitute the sole instrument common to demonstration used in every science [and so] logic is the study of true knowledge" (Piaget, 1949, pp. 2, 4). Since Piaget could not accept the conclusion drawn by logicists, that mathematical induction is reducible to logic, Piaget could have sided with Poincaré by assigning equal importance to both intuition and logic in mathematical reasoning. But this would not do either. Poincaré regarded intuition as analogous to perception which occurred as a "sudden illumination within the mind". Piaget (1974/1977) did not accept this view. How could he? It was during the 1930s that his commitment to a view about the basis of knowledge in action was elaborated in his account of infant development. And this leads to a second influence.

There was a second influence on Piaget's account of infant development. Piaget (1937/1954, p. 188) followed Poincaré in requiring group structures for objective experience. There was, however, an important qualification. What had to be reckoned with as well, Piaget (p. 319) would argue, is vertical décalage manifest as the time lag between the presence of a group structure in practical intelligence by the end of infancy and the later emergence of a group structure in understanding during late childhood. This décalage left open the working hypothesis that the development of knowledge had its basis in action (Piaget, 1941). As Piaget (1936/1953, p. 356; my translation) put this: "intelligence constitutes an organizing activity whose functioning extends that of biological functioning the transcendence of which is due to the elaboration of new structures". If mathematical induction is due to intuition, this takes place "in action". Piaget (1974/1976) later marked this as the advance from practical success to representational understanding.

A third influence is the distinction which Piaget attributed to Poincaré between empirical learning, on the one hand, and necessary knowledge, on the other. A physical system may be such that some property is conserved in that

> someone can know in advance that there will necessarily be the conservation of some property, even though only experience permits the identification of which one it is. (Piaget, 1967a, p. 109)

This is an important distinction in Piaget's work in two ways. First, it fits the development of necessary knowledge which Piaget stated to be his central problem (Smith, 1993, 1999d, 2002a). Second, it fits his research programme directed on the empirical investigation of normative problems (Smith, 1993, 1999b, 2002b).

A fourth influence concerned the terms of reference for Piaget's working hypothesis about the development in reasoning by mathematical induction during childhood. Piaget's proposal was based on his operationalization of intuition in terms of coordinated actions. Recall that Poincaré characterized mathematical induction as reasoning by recurrence. The title of the paper reporting the empirical study due to Inhelder and Piaget (1963) was "From action iteration to elementary recurrence". Their final conclusion was explicitly stated to be against Russell's thesis and "in favour of Poincaré's thesis that reasoning by recurrence would constitute a system of inference specific and unique to numerical realities" (Inhelder & Piaget, 1963, p. 117; cf. p. 118). This conclusion is spectacularly optimistic. It concerns a "specific and unique" form of inference and so this is a commitment to domain-specific reasoning; it refers to intellectual development during childhood, not adolescence; it deals with sound reasoning in advance of logical deduction; and this reasoning is also modal and so understood to be necessary (Inhelder & Piaget, 1963, p. 115).

Inhelder and Piaget's (1963) study had a precursor in their better known study of the conservation of area (Piaget *et al.*, 1948/1960, Ch. 11). Central to the latter was the Euclidean axiom: *if equals are subtracted from equals, the wholes are equal* (cf. Heath, 1956). In the study of geometrical reasoning, the children were presented with two rectangular spaces (fields) which were accepted to be equal in size. Located in each field were the farmer's buildings which were spatially arranged in the same way. The children also accepted that the size of these buildings was the same in each field. This initial array thus conformed to Euclid's axiom. In other words, there was an equality. The buildings in one field, but not the other, were then spatially transformed. The children were afterwards asked whether the fields were still the same. The answer "yes" supported by the child's explanation was interpreted as deductive reasoning in line with the Euclidean axiom. By contrast, non-conserving responses were interpreted as non-deductive inferences. Their main conclusion was that mathematical reasoning develops from "induction which is empirical and intuitive to operational generalization which is deductive" (Piaget *et al.*, 1948/1960, pp. 221, 225).

This conclusion is ambiguous. A natural reading is that the conclusion concerns the passage from induction to deduction. But empirical induction and mathematical induction are not the same thing (see Chapter 2). An empirical induction is an inference from particular to general; logical deduction is the converse from general to particular. Logical deduction is due to a necessitating relation, entailment. Empirical induction is based on a relation of probability. These differences are exclusive. How then could one be developed from the other? A different reading of the conclusion is in terms of reasoning by mathematical induction as a mediator between reasoning by empirical induction and by logical deduction. This is because mathematical induction shares both of the exclusive properties without being equivalent to either.

3.3 Inhelder and Piaget's (1963) Study

As far as I can tell, commentary in English on Inhelder and Piaget's (1963) study has been confined to exegesis by Flavell (1970) and nothing from anyone else. But this study is important on two counts. First, it deals with children's reasoning by mathematical induction in a research context where there are no rivals. This point will be defended in Chapter 4. Second, it contains an interpretation of children's reasoning which is definitely contrary to the "standard" interpretation of Piaget's position (see Lourenço & Machado, 1996). In view of the unavailability of an English translation, a fuller review is now presented.

The ambiguity of the conclusion drawn by Piaget *et al.* (1948/1960) was clarified when Inhelder and Piaget (1963, p. 48) stated their case by reference to a distinction:

$$(A = B) \rightarrow (A_n = B_n)$$

$$[(A, B) = x] \rightarrow [(A, B)_n = x]$$

The former can be read as stating that a numerical equality entails a corresponding equality whatever the number may be. The latter can be read as stating that an equality in some respect entails an equality in the same respect whatever the number may be. This symbolic distinction is struggling hard to make a good point which could be central to the mediation played by mathematical induction in children's reasoning.

The struggling distinction is palpable. *Prima facie*, there appears to be a distinction without a difference. Variables are posited with neither an interpretation nor rules of inference. It is not clear whether the logic is extensional ("→" means "if . . . then") or intensional ("→" means "entailment"). Thus it is not clear how this distinction is to be understood at all. Frankly, there are precedents for this negative reading of Piaget's logic (Braine & Rumain, 1983; Parsons, 1960). Equally, there is a positive counter-case (Apostel, 1982; Smith, 1987; Mays, 1992). And this leads to the good point.

The good point is also evident. One interpretation of the good point is that Inhelder and Piaget are distinguishing the use of knowledge and the development of knowledge. This is a distinction which Inhelder (1954/2001; Inhelder & Piaget, 1979/1980) was especially keen to promote in the demarcation of "her" functional from his "structural" perspective in "their" interdependent work. In the 1963 study, their position ran thus. The first formula

$(A = B) \rightarrow (A_n = B_n)$

fits logical deduction on the basis of substitution in a deductive algorithm. This does not require recurrence (mathematical induction) since an available algorithm is put to use. This use may be novel. All the same, its availability is assured. And this leaves unresolved the question of how access to the algorithm was gained in the first place. Further, such use requires the conservation of number. How did this construction take place (Inhelder & Piaget, 1963, p. 110)? By contrast, the second formula

$[(A, B) = x] \rightarrow [(A, B)_n = x]$

fits recurrent reasoning in that equality (similarity) in some respect is generative of a true equality in the inferential act. Their case runs like this. There is more to reality than the actual world of physical objects (Piaget, 1937/1954). Abstract objects are in the reckoning as well, and numbers are paradigm cases of abstract objects (Inhelder & Piaget, 1979/1980). Any reference to abstract objects runs into the twin problems noticed in Chapter 2, one in ontology and the other in epistemology. It is the latter (epistemological) problem which is central to Piaget's model where the key question is "How does the construction of number take place?" Notice that this construction is more than an empirical induction. It is notorious that empirical induction is fallible (Russell, 1912/1959). A predicate such as "green", ostensibly descriptive of green objects, is always such that some other predicate "grue", applied to objects green up to today and blue tomorrow, could also fit the same historical range of cases (Goodman, 1979). But according to Piaget (1942; cf. Benacerraf, 1965), each number is a distinct class. What requires explanation in ontogenesis is how children construct these new numbers on the basis of their knowledge of the numbers already known to them. Granted: preschoolers have a principled knowledge of small numbers (Gelman, 1972, 1997). How then do they generalize their knowledge of "some" numbers to "other" numbers, even to "all" numbers? At issue here is not the psychological question about how the transfer of knowledge takes place over all contexts and domains, important though this is (Demetriou, 1998; Demetriou & Kazi, 2001; Donaldson, 1992; Hughes, 2000). What is at issue is the equally important epistemological question about understanding a universal property of an object in any one context or domain (Smith, 1999b). Exactly the latter, though not the former, is required for an understanding of the conclusion of an inference due to mathematical induction. This is because such an inference runs from what is shown to be true of a specified number n (*n'importe quel nombre*) and what in that case is also shown to be true of its successor $(n + 1)$ to a conclusion about any number n whatsoever (*nombre quelconque*). It is just such an inference which is at issue in mathematical induction and it was Inhelder and Piaget's (1963) contention that their evidence showed that children were capable of this type of reasoning (see also Piaget, 1942, p. 242; 1967b, p. 412; 1977/1986, p. 303).

Inhelder and Piaget (1963) designed five studies in their investigation of children's reasoning by mathematical induction. Two of these are illustrative. At the outset of study I, the children's familiarity with the words to be used in the questions was checked. Each child was then shown two piles of counters which differed in colour.

Two empty, transparent glasses were also available. The child was asked to take a counter from each pile, one in each hand, and at the same time to place one in one glass and the other in the other. This action was then repeated, leading to three questions as follows. (a) After several iterations, the child was asked whether the two collections were equal. (b) After further iterations, the glasses were screened and the question was repeated. (c) After more iterations, the same question was asked about anticipated additions *all afternoon* or *for a long time*. In study II, the procedure was the same except that a counter was added to one glass with the other glass remaining empty at the outset before any additions were made. This initial inequality was clearly visible to, and detected by, the child. The child was then asked to make equal additions in the same way as in study I. Evidently, these studies map on to two Euclidean axioms relevant, namely *if equals are added to equals, the wholes are equal* (additive equality) in study I, and *if equals are added to unequals, the wholes are unequal* (additive inequality) in study II (cf. Heath, 1956).

In study I, quantitative findings based on a sample (*n* = 300) of children aged 5–7 years were presented in line with three levels of response (see Table 3.1). Two findings are clear. One is that well over half (60%) of the youngest children succeeded on this task at the highest level C. The other is that almost all (+90%) of the older children were successful at level C. Although it is tempting to report this as an age-related increase in children, it is more to the point to relate this improvement to the three levels of response (Smith, 2002a). This is because the epistemic subject (level of knowing) and the psychological subject (knower or individual child) are not the same thing.

The three levels of response can be explained through the qualitative evidence (see Box 3.1). Only a minority (5%) of the responses of the youngest children (5 years) were at level A, where initially sound reasoning lapsed into unsound reasoning. Further, the recapitulative *always* is a retrospective summary of completed actions. The first example in Box 3.1 is illustrative (Inhelder & Piaget, 1963, p. 60). About one third (35%) of the responses of the youngest children (5 years) were at level B, where reasoning switched to and fro between sound and unsound reasoning. The second example in Box 3.1 is illustrative (p. 63). Nearly two thirds (60%) of the

Table 3.1: Percentage of children (*n* = 300) reasoning by age and level in Inhelder and Piaget's study I.

Age	Level		
	A	B	C
5	5	35	60
6	0	7	93
7	0	9	91

Source: Inhelder & Piaget (1963, p. 68)

Box 3.1: Examples of children's reasoning at levels A, B, C

Level A reasoning

Interviewer	Child: Gras (5;1)
. . .	It's the same.
Why?	Have put two at once.
[The glasses are hidden]	It will be the same. [Continues]
Will the glasses be full at the same time?	I don't know.
Why?	There's more there, I don't know . . . I have always done it like that (gestures).

Level B reasoning

Interviewer	Child: Cour (6;2)
Count them.	Sixteen.
And the others.	I know. [Continues]
And now?	I can't be sure.
Why?	Because you can't be sure [Continues]
And now?	The two the same.
Why?	Because you have put 1 and 1.
You said you couldn't know.	Only sometimes.
But can you know if it's the same number?	Yes, you can.

Level C reasoning

Interviewer	Child: Stion (5;9)
. . .	They're the same.
[Screened after the eighth addition]	There's no need to hide them, I know. If I had once taken two, that would be different, but I never took more than one in one hand.
And if you continued for a long time, all afternoon?	You would have to have such big glasses, there aren't any so big.
But can you imagine that?	You couldn't know how many there are.
OK, but can you know if there is the same or not?	Oh! Yes — once you know, you always know.

Source: Inhelder & Piaget (1963, pp. 60, 63, 66)

responses of the youngest children (5 years) and almost all (+90%) of the responses of the older children (6 and 7 years) were at level C, where sound reasoning resulted in generalization. Further, the iterative and almost recurrent use of *always* is prospective and generalizing. The final example in Box 3.1 is illustrative (p. 66).

Four general points can be made about this evidence. First, Inhelder and Piaget simply assumed that their unit of analysis is familiar, namely that their concern is with the quality (level) of reasoning, not the child. At issue are the respects in which reasoning at these three levels is different, not the respects in which these children are different. What is "at a level" is thinking, not the thinker (Smith, 1993, sect. 18). This difference is non-trivial because, second, level C is defined in terms of generalization (universalization) about "all numbers". It is for this reason that Inhelder and Piaget (1963, p. 49) stressed the understanding of what is true of *nombre quelconque*, true of any number at all. What is not at issue is whether the children can transfer their knowledge, for example to other tasks. Third, the word *always* is used at both level A and C. According to Inhelder and Piaget (1963, p. 65), this word is ambiguous in this context. It is apposite at this point to recall that Frege envisaged logic to be a rational means to dispel confusion due to language (see Chapter 2). No doubt it is for this reason that Inhelder and Piaget requested children to give their reasons for their responses which would otherwise remain ambiguous. Fourth, the level C remark made by Stion, *once you know, you know for ever*, is arresting. It is testimony to a realization that counting is not the only way to gain numerical knowledge which can instead be due to reasoning. Piaget (1967b, p. 412; 1977/1986, p. 303. See also Piaget & Garcia, 1987/1991, p. 38 but note the incorrect reference given by these authors on this page) was impressed with this remark which he continued to cite in his later work. Fifth, levels B and C are marked by the use of modal language. In both, the modality is instrumental, dealing with what *can* or *cannot* be known. Further, it is epistemic rather than alethic, since it concerns certainty–uncertainty. This is different from the necessity–impossibility of conclusion on the basis of its premises (Smith, 1997). There is, however, a significant omission concerning children's modal knowledge. Inhelder and Piaget's analysis of their own data is virtually silent about the modal knowledge displayed in their own protocols other than a final remark in their conclusion.

In their interpretation of this evidence, Inhelder and Piaget drew attention to a distinction between (quantified) conservation and (logical) transmission:

> in all of the cases in which we have previously studied the constitution of notions of conservation (conservation of sets or classes, number, length, surface, volume, physical quantity, etc.), we have spoken of the "conservation" of a property through modification of an object or collection only when that property belonged to a completed totality (object or set) merely by transforming its internal relations (changes in layout, or distribution, etc., leaving the amount invariant). We shall say by contrast that there is "transmission" (and not simply conservation) of a property when the latter characterizes first of all a totality *A* and is found again unchanged in a new totality *B*, if the totality *A* is transformed into *B*. In that case, there is no longer only invariance with respect to internal

modifications but also with respect to the construction of a new object or new collection. We shall therefore speak of a causal "transmission" if an object is modified from the outside, or a recurrent "transmission" if a property φ passes from n to $n + 1$. (Inhelder & Piaget, 1963, pp. 107–108)

On this interpretation, mathematical induction requires transmission rather than conservation. In the case of arithmetic, a property true of any (*n'importe quel*) particular number is reckoned to be transmitted to any (*quelconque*) number at all (p. 49). Further, it is contended that reasoning by induction in arithmetic contributes to the construction of number along with deductive reasoning based on conservation. Although both are inferential, the implication is that neither is reducible to the other.

Evidently, three claims are being made here. The first claim is that Piaget's (1950) theory is epistemological, i.e. it is a theory of knowledge. A standard way of gaining knowledge is through reasoning, including reasoning by mathematical induction. Piaget and Inhelder (1966/1971, p. xiii) identified two accounts of knowledge, one for rejection and the other for retention. They rejected the view of knowledge as copy on empirical and theoretical grounds (Piaget & Inhelder, 1966/1971, pp. 385–386). Their position is summarized in the demonstration that a *unitary* object of knowledge, such as the series of integers, is in fact represented in *different* ways, such as a staircase, zigzags, etc. (p. 380). But if there is a many–one relation between representation and object, this means that objects of representation and objective knowledge are not the same thing (Smith, 1999f; see also Bickhard, 2002; Bickhard & Terveen, 1995; Müller *et al.*, 1998).

The second claim is that Piaget and Inhelder retained the view of knowledge as assimilation. The central principle of this view is that knowledge is always based on assimilating action directed on objects. "Knowing the object means acting upon it in order to transform it, and discovering its properties through its transformations. The aim is always to get at the object. Knowledge is not however based only on the object, but also on exchange or interaction between subject and object resulting from action or reaction of the two" (Piaget & Inhelder, 1966/1971, p. 387). The distinction between recapitulative–anticipatory uses of language is comparable to their distinction between reproductive and anticipatory imagery (p. 6). This means that construction is fallible both as to which objects are constructed and which properties are attributed to them (Haack, 1978). The consequence for ontogenesis is that children have to construct all objects as a prerequisite of the discovery of their properties. Note that this construction has a social element.

The individualization of knowledge does *not* mean that individuals construct knowledge "by themselves" since socio-cultural transmission is necessary, but not sufficient, for construction (Smith, 1996b; Mays & Smith, 2001). During childhood, formal reasoning is highly contextualized in a non-reversible system through its mediation by concrete support (Inhelder & Piaget, 1955/1958, p. 216; cf. Adey & Shayer, 1994). Despite the fact that commentators such as Chapman (1992) and Klahr (1999) have found Piaget's (1975/1985) model of construction to be obscure, its motivation is clear enough and it fits Inhelder and Piaget's (1963) study (for discussion, see Smith, 2002a).

This leads to the third claim. The children can easily perform the physical action of jointly adding one counter to each container and they can also repeat this action. This action is repeatable (iterative), resulting in a serial addition to the physical objects, namely the containers. But numbers are abstract objects with their own properties, notably the property of iteration in the number system. Inductive inference from actions and physical objects to their properties is empirical. By contrast, inductive inference from abstract objects to their properties is mathematical. Yet such reasoning is possible only if these abstract objects are constructed by the reasoner in the first place. In short, construction due to assimilation is constraining and enabling. Constraint occurs due to the potential confusion of one set of properties (action on physical objects) with another (operation on abstract objects). Children have constructed the physical world well beforehand during infancy. The task of childhood is to construct other abstract objects, such as mind, society and value — and of course number. Enabling construction is the "creation of new objects (*êtres*) such as classes, numbers, morphisms etc." (Inhelder & Piaget, 1979/1980, p. 21).

3.4 Conclusion

In sum, two main conclusions were drawn by Inhelder and Piaget (1963, p. 115):

> [mathematical induction amounts to] a recurrence which is elementary or, as we would rather say, co-extensive with the construction of number. In fact, what is discovered by children who generalize initially observed equalities is not only that these equalities are true of 1 or of *n*, but especially that if they are true of *n*, they are still necessarily true of (*n* + 1).

One claim is inferential: the children infer a true conclusion on the basis of iterative actions on actual objects which contribute to the construction of the iterative properties of number as an abstract object (cf. Inhelder & Piaget, 1979/1980, p. 108). The other claim is modal: the children display their modal knowledge about the necessity of the inference in that what is true of one number must be true of any number at all (Inhelder & Piaget, 1963, p. 112). Both claims are distinctive. If Poincaré was right to insist that reasoning by mathematical induction required both intuition and logic, Inhelder and Piaget have reconceptualized his position by requiring there to be some other epistemic instrument which contributes in advance of, or along with, deductive reasoning. But in their view, this instrument is not the intuition of a whole series of numbers. Rather it is the recurrent property of action in as much as "this iteration constitutes precisely the elementary instrument of transmission" (p. 69). It is in this sense that "the general coordination of actions" (Piaget, 1967/1971, p. 345; 1977/1995, p. 89) is the basis of intellectual advance.

Their conclusion is significant on two counts. First, it concerns a major type of reasoning which is mathematical induction. Second, their evidence was interpreted through strong claims about the onset of such reasoning during childhood. Neither has been given adequate attention in research in psychology and education.

Chapter 4

Research on the Development of Children's Reasoning

The aim of this chapter is deliberately limited as a review of research on human reasoning with special attention to its relevance to the development of children's reasoning by mathematical induction. The main conclusion is that Inhelder and Piaget's (1963) study is virtually unique and it is for this reason that their study is the basis of the empirical project in Chapter 6.

In psychology, the terms *reasoning* and *thinking* are used sometimes interdependently (Garnham & Oakhill, 1994), and sometimes severally with an implied reference to the other (Siegler, 1978). It has even been claimed that "thinking, broadly defined, is nearly all of psychology; narrowly defined, it seems to be none of it" (Oden, 1987, p. 203). Two questions about reasoning are notorious trouble-spots, one about language (Fodor, 1975) and the other about consciousness (Carruthers, 1996). Neither is open to easy resolution (Holyoak & Spellman, 1993). Both are ignored here.

4.1 Reasoning

As a working definition, Johnson-Laird's (1999, p. 110) proposal is exemplary:

> reasoning is a process of thought that yields a conclusion from percepts, thoughts, or assertions.

There are two good points to make about this definition along with one limitation. One good point made by this definition is that it makes clear what is common to all types of reasoning. What is common to all types of reasoning is some relation. This relation links a conclusion (one thing) with something else, i.e. the relation has two terms. According to Johnson-Laird, one term covers percepts, thoughts or assertions. The other term is the conclusion drawn from these. Drawing a conclusion is making an inference. As such, it is a characteristic and important type of human activity. It is also something which is, on the one hand, easy to do, and yet also very hard to do well. This means that there is ample scope for differences. An important class of differences in reasoning is due to differences in the relational link. Thus deductive reasoning is based on one type of relation, and inductive reasoning on another. Indeed, the dispute between Poincaré, Frege and Russell which was reviewed in Chapter 2 was about exactly what this relation is in the case of mathematical induction. There are two ways in which this relational link can be codified. One is through normative logic. Logic sets out the formal properties of the different relations in deduction, induction, and so on (Haack, 1978; Sainsbury, 1991). The other way is through empirical psychology. Psychology sets out the properties of actual reasoning in specified contexts (Braine & O'Brien, 1998).

A further reason why Johnson-Laird's definition is instructive is due to its inclusivity. The definition covers inferences from percepts and thoughts, and not merely assertions, unlike a more restrictive definition in logic texts in which the former pair would be excluded (cf. Quine, 1972). Thus the definition is compatible both the formal study of reasoning in logic, and with the empirical study of reasoning in psychology.

But this definition is incomplete due to the exclusion of action. The definition makes a good fit with theoretical reasoning in which the conclusion is a thought. This has traditionally been the main focus in logic (Sainsbury, 1991). But there is more to logic than this. In practical reasoning, the conclusion is an action which according to von Wright (1983a) is derived from beliefs and desires on the basis of deontic logic. In Piéraut-Le Bonniec's (1980) model, modal reasoning has its origin in children's actions about what is makable–unmakable. So the definition of reasoning needs to be widened to ensure that action, and not merely thought, is covered.

The review now looks at research on three types of human reasoning, deduction, induction, and mathematical induction, examples of which were given at the outset of Chapter 2.

4.2 Deduction

The central relation which links the premises of a deductive argument to their conclusion is entailment (Hintikka, 1996, p. 4). It is this relation which ensures that all deductive arguments are valid arguments in that the premises necessitate the conclusion. Here are some typical definitions:

> [In a valid argument] the truth of the premises is *inconsistent* with the falsity of the conclusion. (Strawson, 1952, p. 13)

> [A logically valid implication is such that] if one statement is to be held as true, each statement implied by it *must* also be held as true. (Quine, 1972, p. 4)

> [An argument] is valid only if it *couldn't* have, not just doesn't have, true premises and false conclusion. (Haack, 1978, p. 22)

> Valid arguments are *necessarily* truth-preserving. (Sainsbury, 1991, p. 15)

There are two important points to notice here. One is that a valid argument is truth-preserving. In a valid argument, premises never could result in a false conclusion, always provided the premises are themselves true. But an argument can be valid, even if its premises and conclusion are false (Smith, 1993, p. 22). Sainsbury (1991) has marked this point by contrasting truth-preservation and truth-production in that a valid argument ensures the former but not the latter. A valid argument whose conclusion is true is a sound argument. The second point is that all of these definitions of validity invoke the modal concept of necessitation (Smith, 1993, 1997). Anything which is *inconsistent* is a contradiction and so is impossible. The term *must* is a vernacular expression of necessity. The modal aspects of *couldn't* and *necessarily* are explicitly modal. Entailment is a paradigm example of a logically necessary relation. These definitions lead to a uniform conclusion, namely that (deductive) reasoning is valid only if it has a modal component. That's because if it is not modal, it is not necessitating and so is not valid either. The point at issue is not that all reasoning is modal (since propositional reasoning is not modal), but is rather that all *valid* reasoning is modal (for example, deductively valid reasoning in propositional logic). This point has psychological implications which will be reviewed below.

Johnson-Laird's (1999) critical review of research on deductive reasoning would be hard to improve. Even so, it is silent about reasoning by mathematical induction. Using his review, available research on deductive reasoning is incomplete in two respects. First, it is incomplete through a failure to discuss the extent of the contribution made by deductive reasoning to mathematical induction. This was the problem in Chapter 2 above. Second, it is silent about the extent to which reasoning by mathematical induction is under development during childhood. This question will be revisited below in the review of research on mathematical induction.

Research in three neighbouring areas can also be noticed here due to their focus on modal reasoning about necessity and possibility.

One is the research programme due to Piéraut-Le Bonniec (1980). One of her central claims had its basis in the work of Inhelder and Piaget (1955/1958, p. 246) who argued that children's understanding of modality is initially an assimilation to their actions which are not yet reversible operations. Piéraut-Le Bonniec took this to be a good point since her own study of children's modal reasoning was based on this claim, i.e. coherence as used by children in the organization of their actions (pp. 52, 143). Thus her research required the broadened version of Johnson-Laird's (1999) definition, which includes action. Second, it was her argument that the logical model of reversible operations used by Inhelder and Piaget is based on extensional (and so non-modal) logic, when a modal (and so non-extensional) logic is required (the difference between extensional and modal logic in developmental theory is discussed by Kargopoulos & Demetriou, 1998; Smith, 1998b). Although Piaget (1949) was aware of the distinction between material (extensional) implication and logical (intensional) implication, this distinction becomes problematic in "natural logic" (Piéraut-Le Bonniec, 1980, pp. 55, 58, 143). Piéraut-Le Bonniec's point is well taken, but it is too strong. Modal logic is not required to understand the modality of validity, i.e. a modal model is sufficient but not necessary. An Aristotelian square of opposition in terms of modal concepts — not operators — was invoked by Piéraut-Le Bonniec (1980, p. 6) and this may be sufficient. Third, her own modal model is a complex system based on alethic, deontic and epistemic modalities (Piéraut-Le Bonniec, 1980, p. 147). This elegant and rich model merits attention both at rational and empirical levels. There are analogies and disanalogies between the different modalities which are discussed independently of evidence about ontogeny (von Wright, 1983a). There are also empirical studies of modal reasoning during childhood without regard for these rational distinctions (Byrnes & Beilin, 1991; Miller, 1986). Piéraut-Le Bonniec's model is in a class on its own in both respects. The direct application of her model to reasoning by mathematical induction is beyond the scope of this study, but the implications of her model for the interpretation of the findings will be taken up in Chapter 8.

A second area is the model of entailment logic due to Piaget and Garcia (1987/1991). Entailment logic is the logic of relevance and necessity (Anderson & Belnap, 1974). This logic is a modal logic which differs from a modal logic based on strict implication (e.g. Lewis & Langford, 1959). Entailment logic avoids the paradoxes of implication (von Wright, 1957). But it also generates paradoxes of its own (Haack, 1978). As used by logicians, both entailment logic and strict implication are intensional logics of deductive reasoning. As used by Piaget and Garcia (1987/1991), entailment logic is a logic of children's reasoning, including both deductive and non-deductive reasoning. In some of their empirical studies (e.g. Chapters 1 and 7), non-deductive reasoning includes reasoning by empirical induction. In other studies, reasoning by mathematical induction is implicit in the design and explicit in the interpretation. In Chapter 3, children were given geometrical shapes (triangles, squares, pentagons, hexagons) and requested to make particular patterns with them. At issue were the inferences that the children made on the basis of their success in combining different shapes. One inference concerned the extent to which this could be continued. The inferences made in level I reasoning included claims about the possibility of only limited continuation, that is "without indefinite iteration analogous to the succession of natural numbers: $n \rightarrow (n + 1)$. The reason for such a limitation is obviously that the recurrence (the 'always') is not observable and that the subject still reasons or infers only with respect to a universe of empirical objects" (Piaget & Garcia, 1987/1991, p. 38). Immediately following is a reference to Piaget and Inhelder (1941/1974), and this reference is false. It should be to Inhelder and Piaget (1963). Even so, reasoning by mathematical induction drops out of account in the interpretation of this study. Similarly in Chapter 4, the task was to combine cardinal and ordinal numbers in an integrated sequence by reference to two boxes linked by a tube. One box contained ten marbles whilst the other box was empty. The children were asked how many marbles would be in the other box before the fifth marble was transferred. Thus a question about cardinality required an explanation in terms of ordinality. In their interpretation of this study, Piaget and Garcia (1987/1991, p. 51) advert to the conservation of number in two systems in that each member of the system is bound by its predecessor, namely $n - 1$. Mathematical induction is similar which is no doubt the reason why Piaget and Garcia (p. 53; note that the standard translation requires modification on this point) conclude with the comment about "a system of inferences about any at all" (*un système quelconque d'inférences*). But an allusion to mathematical induction in relation to a study of the combined use of two number systems is not the same as a study of mathematical induction.

A third area concerns the development of proof construction in studies, First, by Foltz *et al.* (1995) in relation to early adolescence, and then by Ricco (1997) in relation to late childhood and adolescence. The main issue in both studies was the understanding of conditionship relations as an element in constructing a valid proof. It is significant that these studies had their basis in Piaget's (1983/1987) study of necessity which was devoid of a model in modal logic. This is both because a modal model was used later by Piaget and Garcia (1987/1991) with the express purpose of "cleaning up" Piaget's logic, and because their model is entailment logic which Ricco (1993) has endorsed. Yet it is apparent that the sequence envisaged in these studies is "a developmental progression from inductive to deductive approaches to the construction of logical proofs" (Foltz *et al.*, 1995). This is comparable to Piaget *et al.*'s conclusion (1948/1960), which was argued in Chapter 4 to be not the same as the conclusion drawn later by Inhelder and Piaget (1963).

In short, available research on deductive reasoning is largely silent about children's reasoning by mathematical induction. This provides confirmation of the importance of Inhelder and Piaget's (1963) study.

4.3 Induction

In deduction, premises provide rationally conclusive grounds for the conclusion. But what happens if the premises present some — but not conclusive — grounds for the conclusion? An inductive argument is such that its premises make its conclusion more probable, i.e. if the premises are true, then the conclusion is more likely to be true than false. In other words, induction is ampliative in that its conclusion "goes beyond" its premises, i.e. the inference is from particular to general (Skyrms, 1995). This means that induction is the classic form of generalization.

Psychological research on inductive reasoning has been reviewed by Garnham and Oakhill (1994). Their review leads to the same conclusion as the previous review of psychological research on deductive reasoning, namely its silence about children's reasoning by mathematical induction. Even so, there are some pointers here which should be noticed. A classic study was reckoned to support the proposal that fallacies of inductive logic are endemic in the human mind with an inadequate grasp of inductive principles (Tversky & Kahneman, 1974). Cohen (1986) argued that, ironically, any conclusion about inherent irrationality of the mind is an overgeneralization of "merely" psychological evidence. A re-analysis of the original studies leading to a new account of the psychological representation of normative heuristics is due to Gigerenzer (1998). But his proposal leaves unanswered the question of how these heuristics of inductive logic are developed by human minds in the first place. It is this question which is central to research on inductive reasoning taken in the wide sense of scientific thinking. Three questions have been dominant.

One is a demarcation question, namely how the distinction between inductive and deductive reasoning is drawn. Both Piaget (1924/1928) and Vygotsky (1994, p. 229) remarked on children's conflation of empirical (causal) with logical (necessary) relationships. The implication is that children have an incomplete, or even defective, grasp of both types of reasoning pending the development of homogeneous structural changes in their system of thought. According to Keil and Lockhart (1999), this commitment to structural change in thought is itself problematic and it is their argument that a more promising position requires heterogeneity of knowledge structures rooted in individual concepts. In reply, it could be argued that this solves one problem by postponing another. In view of the difference between reasoning by induction and deduction, the use of one concept in one knowledge structure (*KS1*), but not another (*KS2*), could account for differences in successful reasoning due to the content difference between *KS1* and *KS2*. Yet the same forms of (inductive, deductive) reasoning fit both *KS1* and *KS2*, and so what is still awaited is some capacity to exploit this formal similarity. Any argument based solely on heterogeneity breaks down (Flavell, 1982).

A second question concerns the contribution made by children *qua* intuitive scientists to their conceptual change. According to Carey (1985), this contribution is substantial and also successful in as much as children can and do restructure their thinking in the development of novel knowledge. Carey (1999) has noted that a fuller elaboration of the intellectual mechanisms responsible for shifts in thinking is awaited. Even so, her account is open to the challenge due to Kuhn (1989) on the grounds that a metaphor is not an explanation and that the analogy between the thinking of the adult scientist and child savant breaks down. The reason for this breakdown is because children's thinking differs from adult thinking in epistemologically important ways (Kuhn, 1997).

This leads to the third question, namely how well children draw the distinction between theory and evidence. Kuhn *et al.* (1988) argued that children do not draw this distinction at all well. There are consequential constraints on their capacity to coordinate these two elements of scientific reasoning. This argument has the implication that the strategies used by children both in distinguishing and in coordinating evidence with theory would have to be powerful enough to cover different domains and contexts (Kuhn *et al.*, 1992). The counterclaim has been made that Kuhn has overgeneralized her evidence in that children can in certain contexts distinguish evidence from theory (Leach, 1999; Ruffman *et al.*, 1993). In turn, this leads to questions about the epistemological principles at work in the human mind during ontogenesis (Hofer & Pintrich, 1997).

Three general conclusions can be drawn from research on reasoning by empirical induction. One is the extent to which children reason by *empirical* induction about number. Another concerns reasoning by mathematical induction on the part of children or adolescents who can reason by empirical induction. Neither has been pursued. A third conclusion is that if there are domains of knowledge, mathematics would count as a domain. Research on reasoning by empirical induction has passed over the opportunity of investigating the distinct form of reasoning by mathematical induction in this paradigm case of a distinct domain.

4.4 Mathematical Induction

The central relation in mathematical induction is such that "the premises of the argument necessarily imply the generalization that is its conclusion" (Skyrms, 1995).

Recall from Chapter 2 that it is the joint presence of these two properties — necessitation and generalization — which makes it such a potent type of reasoning.

Although there is much research on the development of children's mathematical reasoning through deductive reasoning, this research has typically by-passed Inhelder and Piaget's (1963) study. Further, it has also by-passed reasoning by mathematical induction generally (Bideaud, 1992; Dehaene, 1997; Gelman & Williams, 1998; Hughes, 2000; Nunes & Bryant, 1996, 1997). It is for this reason that the survey which now follows is selective in dealing with salient themes in available research.

An account of the development of numerical knowledge during early childhood is due to Gelman and her colleagues. It is reckoned to be a rival to Piaget's account. The central claims are threefold. One is that number development has its basis in preschoolers' cognitive competence (Gelman, 1978). This first claim is widely accepted (Nunes & Bryant, 1996). A second claim is that this competence has a principled basis in five counting principles (Gelman & Gallistel, 1978; Gelman *et al.*, 1986). This claim has been contested in that not all abilities amount to principled knowledge (Baroody, 1992; Fuson *et al.*, 1988). Her third claim is that *number knowledge* is the principal element in this domain, manifest well in advance of *number reasoning*. In this regard, Gelman (1997, p. 309; cf. Gelman & Gallistel, 1978, pp. 64, 239) explicitly denied that evidence about preschoolers' number knowledge is evidence about preschoolers' number reasoning based on conservation. This is an important denial. If counting is in advance of conservation, this means that they are not the same thing. Indeed counting and conservation are independent ways of gaining knowledge in mathematics (Sophian, 1995). This was acknowledged by Gelman and Gallistel (1978, p. 201) in their reminder that counting is not required for certain forms of number knowledge, for example that there is one rider for every horse in a troop of cavalry (cf. the waiter argument in Frege, 1884/1950, p. 82; see Chapter 2 above). Crucially, to be primary is not thereby to be a precursor. This is because a predecessor can constrain its successors, unlike a precursor which contributes to its successors. A case in point is due to principled differences in different aspects of number. For example, each natural number has a unique successor, unlike the infinity of rational numbers between two fractions (Hartnett & Gelman, 1997, p. 343). Granted: Gelman's account has rightly made the point that number development is manifest throughout childhood in that preschoolers have a principled capacity to count. Much less clear is how this capacity interacts with number reasoning. There is an evident ambiguity in the notion of one–one matching, which concerns the relation of number–word and object in Gelman's account but which concerns one and the same number in two sets of objects in Piaget's account (Bryant, 1995). Crucially, Gelman's work has not squarely addressed the question of how the capacity to count and the capacity to reason combine in children's arithmetic. This is an important question since equality is the central relation in arithmetic, at least on Frege's view quoted in Chapter 3. The fact of the matter is that equality cuts across both knowledge-based counting and reasoning by conservation.

How, then, do children develop an understanding of numerical equality which is required in their getting to know the equivalence of their correctly counting in different contexts? The point at issue can be clarified through a famous argument due to Elkind

(1967). Elkind's argument was that Piaget's analysis of conservation merged two questions, one about identity — i.e. equality; see Chapter 3 — and another about equivalence. But this argument runs into a main defect in that it is question-begging. The inference to identity (equality) on the basis of equivalence won't work. Although identity (equality) entails equivalence, the converse is false (Marcus, 1993). Elkind's argument therefore fails to explain how a logically weaker relation (equivalence) generates a logically stronger relation (identity). Further, this difficulty undermines any account of the development of number knowledge which is divorced from numerical reasoning. This is because numerical reasoning hinges on generalizations from any number n to $n + 1$. And this seems to point in the direction of mathematical reasoning through formal proof, which is well beyond the capacities of children.

Mathematical proof is largely absent from the secondary school curriculum in the United States where it is confined to geometrical proof (Moore, 1994). The same is true of the National Curriculum for Mathematics in Great Britain (DES, 1991; DfEE, 1999). In consequence, there is an abrupt transition to proof in higher-level mathematics courses which "is a source of difficulty for many students who have done superior work with ease in their lower-level mathematics courses" (Moore, 1994, p. 249). Moore's interpretation of this difficulty is in terms of the mismatch between a concept definition and a concept image. The former is mathematical and is manifest as a formal definition of some concept which is valid in mathematics. The latter is psychological and "refers to the set of all mental pictures that one associates with the concept" (p. 252). The difficulty faced by advanced students lies in their reformulation of the latter through the former. This interpretation is comparable to de Saussure's (1960) definition of a sign as a concept-image. It therefore suffers from a fatal weakness since the initial distinction between definition/image is exclusive in that neither is the same as the other. And it is seemingly exhaustive as well. What this distinction fails to indicate — or even to imply — is whether there can be a mediator whose properties intersect both.

One mediator between inductive and deductive reasoning is transformational reasoning which is "the mental or physical enactment of an operation . . . that allows one to envision the transformations that these objects undergo and the set of results of these operations" (Simon, 1996, p. 201). Such reasoning is dynamic in that new cases are generated from the cases available to the reasoning subject. It is enacted either mentally "in the mind" or physically "in action". It is anticipatory, not reproductive. It is pointed out that teachers — and researchers — have paid too little attention to the important role of this form of reasoning with regard to the making of mathematical sense (p. 207). This proposal is attractive, even though the difference between empirical and mathematical induction is not drawn. There is, however, scope for the identification of transformational reasoning in reasoning by mathematical induction.

Another mediator is reported by Flavell (1970, pp. 1003–1006) in his lucid survey of a range of Genevan studies on number development during childhood. These studies straddle the counting–conservation divide. None has yet been translated into English, although some have attracted commentary (Bryant, 1995). Even so, Flavell's (1970) paper contains the sole review in the anglophone world of the study due to Inhelder and Piaget (1963) in which recursive reasoning in arithmetic is based on iterated action.

Piaget (1942, p. 242; 1967b, p. 412; 1977/1986, p. 303) regularly invoked this type of reasoning. The 1963 study is significant in view of the fact that this mediator is identified in the context of the dispute about reasoning by mathematical induction.

Accordingly, it is this study which is the basis of the replication and adaptation in Chapter 6. Prior to that, however, a methodological question has to be addressed in the next chapter.

Chapter 5

Reasoning, Reasons and Responses

The distinction between methodology and method is important. Methodology is a branch of the philosophy of science. It deals with questions about the differences between formal sciences, such as mathematics, and empirical sciences, such as physics (Nagel, 1961); about the aims of scientific research programmes directed on the verification (Carnap, 1962) and falsification (Popper, 1979); about the nature of explanation and understanding in the natural and social sciences (von Wright, 1983b). Method is one element of an empirical study, setting out the design, sample, instrumentation and procedure (Harris, 1986; Keppel, 1991). This means that methodology and method are not the same thing, even though there may be a middle-ground covering both (Brown, 1988, Laudan, 1984). A method always presupposes a methodology. The latter does not have to be set out except in cases where doubts about the soundness and applicability of a method are likely to arise.

Statistical testing serves to bring out the point here. The method of statistical testing is commonplace in psychological research generally (Kinnear & Gray, 1997). An empirical question is posed, usually as a hypothesis about a causal relationship. A study is designed as a test of this hypothesis, usually through quantitative findings. These findings are then described and analysed by means of appropriate statistical techniques. Although this is common practice, this does not mean that the practice is thereby commendable. Not all practices — nor even typical practices — amount to good practice. And this is where methodology fits in. Anyone who sets out to justify or to challenge the legitimacy of a practice is addressing a problem of methodology. Recent challenges to statistical testing in psychology have been expressed on several counts. One challenge is to the misplaced primacy given to inferential testing at the expense of statistical redescription (Wilkinson, 1999). A second challenge is that the basic conditions of measurement and scaling have been massively disregarded in psychological research which amounts to an inappropriate extrapolation from measurement principles in the physical sciences (Badia et al., 1970; Michell, 1997). A third challenge concerns the legitimacy of statistical testing in psychology at all, which is stated to be an "incoherent mishmash [based on] the 'hybrid logic' of statistical inference" (Gigerenzer, 1993, p. 314; cf. Gigerenzer, 1989, 1998). A fourth challenge is the non-empirical contribution to the advancement of any science with the empirical science of psychology identified as being ripe for development due to its preoccupation with the pseudo-empirical (Kukla, 1989; Machado et al., 2000; Smedslund, 1994). This example is a salutary reminder that the question of complying with conventional practice and rationally accounting for that practice are not the same thing.

Just as there are divergent views about the function and limits of quantitative data, there are also divergent views about the fucntion and scope of qualitative data in the social sciences. In educational research, qualitative data has an established place for use along with quantitative data (Erickson, 1992; Hamilton & Parlett, 1977; Light & Pillemer, 1982; Mason, 1996; Maxwell, 1992; Schrag, 1992). In psychological research, priority is generally assigned to quantitative data to the virtual exclusion of qualitative data, and this is expecially marked in developmental psychology (Appelbaum & McCall, 1983; Baltes & Nesselroade, 1979; Weinert & Schneider, 1999). Since joint methods are used in Chapter 6, the purpose of this chapter is to provide a methodological rationale for this stance.

5.1 Critical Method

The particular challenge which is addressed here concerns the use of a critical method (Piaget, 1947; Piaget & Inhelder, 1961). This method is summarized by Inhelder *et al.* (1974; see also Smith, 1993, sect. 11). The use of such a method is recommended by some developmentalists (Chapman, 1988; Ginsburg, 1997). It is complementary to a microgenetic method (Kuhn, 1991; Miller & Coyle, 1999; Siegler, 1996). But it has also been criticized as being especially unsuitable for research on intellectual development during early childhood (Brainerd, 1973; Siegal, 1997). In view of the fact that the children who participated in the study in Chapter 6 were aged 5–7 years, it will be instructive to provide a rationale for the use of a critical method in this study.

A critical method has three elements, namely judgment as the unit of analysis, standardized activity for contextualization, and explication of the judgment's grounds.

5.1.1 *Judgment as the Unit of Analysis*

Both Aristotle (1987) in his *Nicomachean Ethics* and Kant (1948) argued that morality requires that an action is not merely in conformity with a moral rule, but also due to the rule. This is an important distinction in the study of children's reasoning. It is one thing to make a correct (true) response or to express a belief which is true. It is something else again to recognize something as true. Recognition defined in this way amounts to judgment (Frege, 1915/1979, p. 251). This definition has the consequence that a correct response and a true judgment are not the same thing. A response has causes for investigation in psychology. A judgment is manifest in acts of human thinking, and these *acts* always have causes for investigation in psychology. But a judgment is the recognition of something as true, i.e. an acknowledgement of its truth. Thus a judgment also has *grounds* which are reasons. Grounds give the reason or reasons why a judgment is true.

Since logic is the formal science of truth, grounds have logical properties. "The task of logic is to set up laws according to which a judgment is justified by other [judgments], irrespective of whether these are themselves true" (Frege, 1906/1979, p. 175). *Acts* of judgment always have *psychological causes*, whilst the *judgment* in any such act is justifiable in terms of its *logical grounds*.

> The causes which merely give rise to acts of judgment do so in accordance with *psychological* laws (but objective knowledge requires something more). And this is where *epistemology* comes in. *Logic* is concerned only with those grounds of judgment which are truths ... There are laws governing this kind of justification, and to set up these laws of valid inference is the goal of logic. (Frege (1891/1979, p. 3; emphasis added)

Under this view, psychology is necessary, but not sufficient on two counts. One is that psychological causation can lead to error (falsity) and so does not invariable lead to truth (Frege, 1891/1979, p. 2). Further, thinking is marked by individual differences, unlike one and the same truth which is accessible to us all as the same truth (Frege, 1918/1977).

The argument that Piaget had a similar view is elaborated elsewhere (Smith, 1999a,b,c). It is pretty clear that Piaget (1949, p. 4; 1967/1971, p. 35) regarded logic as the formal science of truth and that Piaget's (1923, p. 57; 1950, pp. 13, 30) developmental epistemology can be viewed as a "linking science" or the empirical study of normative facts (Isaacs, 1951; Smith, 2002a,b). Acts of judgment are central to this epistemology which sets out to explain the extent to which normative truths are first instantiated in, and then judged to be true by, human minds.

So interpreted, an act of judgment is an inclusive unit of analysis. Psychology is always required in the causal investigation of mental activity. But something more is required as well to ascertain whether, how and why what is thought is also thought to be true.

It should be pointed out that not all responses, nor even responses based on good reasons, are acts of judgment. It is a requirement that all acts of judgment are based on responses for which an agent has a corresponding reason. An outstanding problem in developmental theory concerns the demarcation and operationalization of this epistemological distinction between correct responses and true acts of judgment.

5.1.2 Standardized Activity for Contextualization

A method is general and from this it follows that its rules and techniques should be used in the same way in different studies. In other words, any method requires standardization. What is less clear is the extent to which standardization should be carried through in any particular study. If an experimental method is used, standardization is complete. By contrast, a critical method requires something to be standardized, leaving other things non-standardized.

The standardized element of a critical method consists in the design of the activity required to "trigger" a response. According to Piaget (1947, p. 7), a critical method requires "the introduction of questions and discussions only at the end, or during the course, of manipulations bearing on objects which arouse a determinate action on the part of the subject". The implication is that, even if such questioning is eventually non-standardized, it is always subsequent to a standardized activity at the outset. This point is important since it means that a critical method is incompletely standardized, not completely non-standardized. At the outset, then, the investigator invites the individual taking part in the study to do something. This requirement is in line with Piaget's commitment to the view of knowledge as assimilation in Chapter 3 and with the broader definition of (practical) reasoning in Chapter 4. As Inhelder (1956; cf Wittgenstein, 1998) put it, *Im Anfang war die Tat* (in the beginning was the deed). Which deed this is to be depends on the design of the task. Many of Piaget's tasks are well known and the original designs have been the subject of massive adaptation (see Smith, 1992). There are three key points to notice here. One is that an action is not a bodily movement but is instead a meaningful activity charged with intentionality (Vonèche, 1999). So construed, an action amounts to the agent's way of making sense of the problem set by the investigator. Second, the same action is compatible with different judgments, i.e. the mapping from action to judgment is one–many. This is exactly the point made by Aristotle and Kant about moral activity and moral judgment. Finally, making a judgment and expressing it are not the same thing. In the case of complex judgments, the sole way to ascertain which judgment has been made is in virtue of its expression. Putting these points together, even though a judgment can be made in a meaningful action without being expressed, an observer can identify which judgment this is only if the judgment is expressed by that agent.

5.1.3 Explication of its Grounds

An *act* of judgment always has causes. A *judgment* always has grounds (reasons). It is a category mistake to confuse a cause with a ground. Even if different causal conditions produce different judgments, that leaves untouched the distinct question "What were the reasons given by the agent for any particular judgment?"

Piaget and Inhelder (1961) addressed this issue head-on by remarking that standardized assessment in experimental studies presupposes a principle of *ne varietur* (let nothing be changed). From this principle, they identified two preconditions of the use of an experimental method, namely:

> we know in advance what we want to get from the child and that we believe we are capable of interpreting the obtained responses. (Piaget & Inhelder, 1961, p. xii)

In their view, these conditions are rarely met in a study directed on intellectual development. First, novel knowledge is inherently unpredictable and could not be specified in advance. Today's child may be tomorrow's genius, or otherwise may make a response which is "new to us". How are we to identify, here and now, all of the novel respects of (all) children's responses? The implication is that an experimental method will artificially exclude phenomena. This is a good point. A scientific hypothesis has been compared by Popper (1963) to a searchlight which may provide powerful illumination of part of a wide stage, leaving the remainder in obscurity. This first precondition is designed to reduce the extent of this obscurity. Second, responses amount to "raw data" and, as such, require interpretation. It is one thing to set out an interpretation by reference to the investigator's view of the world. It is something else again to set out an interpretation by reference to the agent's own reasoning which is undergoing development. But reasoning which is undergoing development may be a prime specimen which fits a prespecified template (this is good reasoning, or it is not). Equally, it may be indeterminate, incomplete, partially defective, non-standard, in short not a standard case at all. But in that case, it poses problems of interpretation. The implication is that an experimental method makes the misplaced assumption that all responses will amount to prime specimens. By contrast, a critical method makes no such assumption, leaving open the extent to which and the respects in which a response is open to interpretation. This is in line with von Wright's (1983a,b) contention that any agent views the world through a distinctive "horizon of intentionality". What matters in interpretation is the identification of the agent's, rather than the investigator's, view of the world.

Both responses and reasons are made available by the use of a critical method which provides a mediating position. Responses are "triggered" by the task. These responses are suitable for statistical analysis with a view to drawing causal conclusions. Reasons given for these responses by the agents who made them are suitable for an epistemological analysis. The aim of such analysis is to ascertain what the reasoning is, not from the investigator's point of view, but rather from the point of view of the person whose reasoning it is. Whether the reasoning is correct or incorrect, whether it is based on logic or pseudo-logic, whether it is rational or irrational, all reasoning is based on a relation between premises and conclusion. This follows from the definition of reasoning given at the outset of Chapter 4. There are of course all manner of ways in which this can be done. Ascertaining how children reason in arithmetic is a central issue in this study.

On methodological grounds, a statistical analysis of responses without the reasons for these responses is only half of the story. If children's responses are based on their reasoning, it follows that two types of analysis are required, one directed on agents' responses and the other on the agents' reasons for these responses. The responses can be pooled for statistical analysis. The reasons can also be pooled for epistemological analysis.

5.2 Rationale for a Critical Method

The use of a critical method is controversial. Particular objection has centred on the requirement that children should give their reasons for their responses. One objection is that this increases the incidence of false negatives, i.e. the non-attribution of knowledge in cases where it is present (Brainerd, 1973). A second objection is that requiring children to give reasons for their responses ignores the fact that children,

Figure 5.1: Questioning children: *When did you last see your father?*

Figure 5.2: Children's questioning: *Daddy, what did you do in the Great War?*

whose knowledge is in place, are likely to be conversational neophytes (Siegal, 1997, 1999). This view is encapsulated in Figure 5.1, where a child during the Civil War in England has been asked a question whose answer is likely to be an answer disadvantageous to the child. Under this view, questioning children is comparable to a grilling, and so to be undertaken sparingly so as to minimize or reduce disadvantage to children.

A countercase is now set out on the basis of five arguments for using a critical method. These arguments concern standardized questioning, the interdependence of "What?" and "Why?" questions, foundational and causal accounts of knowledge, the internality and externality of relations, and the functions of statistical and epistemological analysis. These arguments lead to a different view of children and questioning which is encapsulated in Figure 5.2, where a child during the First World War is asking a good question. Under this view, children's questioning is at issue in that children — and not merely investigators — are co-participants in dialogues. Children have lively minds which are capable of novelty. As such, adults' questioning should be undertaken in such a way as to maximize the advantages to children in virtue of their contributions to the dialogue.

5.2.1 Standardization of Questioning

The standardization of questioning can occur in two different ways. It is one thing for the *question framed by the investigator* to remain self-identical across the sample. It is something else for the *question framed by the individual* to remain self-identical through that person's train of thought. In both an experimental and a critical method, there is a requirement for the questions posed at the outset by the investigator to be standardized, i.e. the same initial questions are put to each child. But it is only in a critical method that there is any control over the extent to which the *question posed by the investigator* matches the *question framed by the individual*. The point is that what is still to be ascertained is:

> how the question is framed in the child's mind, and even whether it is
> framed at all, [so as] to situate each symptom in a mental context instead
> of abstracting it from this context. (Piaget, 1926/1929, p. 7)

It is this key point that lies behind the two conditions set out in the previous section by Inhelder and Piaget (1961), namely that investigators may not know what can be gained from children, still less how to interpret this in every case. One and the same question as standardized by the investigator may be contextualized in the human mind in an indefinite number of ways. In Piaget's constructivist model, this is exactly why context is a powerful variable. Such cases include:

(i)   belief accepted without reasons
(ii)   belief based on irrelevant personal reasons
(iii) belief based on partial reasons

(iv)   belief based on good reasons without recognition of its truth
(v)   true knowledge
(vi)   judgment or the recognition of the truth of what is known.

Notice that each of these is compatible with a psychologically correct response. Yet each of (i)–(vi) is epistemologically different. It is a fallacy to infer any one of these to the exclusion of the others solely on the basis of a psychologically correct response. If Moshman (1994, 1998) is correct in maintaining that reasoning based on reasons leads to reason, an analysis of children's reasons is indispensable in the identification of the type of reasoning at issue. Another way of making this point is by reference to the specific character of intentionality, i.e. the beliefs, desires and actions which are used to make sense of the world. An agent views the world through a "horizon of intentionality", and only thus is it meaningful to that agent (von Wright, 1983a,b). It begs the question for any one agent to prescribe the specific character of someone else's "horizon".

5.2.2 What-questions Depend on Why-questions

In his *Posterior Analytics*, Aristotle (1987) argued that What- and Why-questions are interdependent. This is because an answer to the question "What is that?" presupposes an answer to the question "Why is it so?" Schematically, if the answer to the question "What is that?" is "That is an *x*", this answer presupposes an answer to the related question "Why is that an *x*?" The suggestion is not that an answer to a Why-question should actually be given by anyone who answers a What-question. Rather, the capacity to answer the latter is presupposed by an answer to the former.

Aristotle's argument fits human reasoning, as Johnson-Laird's (1999) definition in Chapter 4 made clear. Reasoning requires there to be a relational link between a conclusion and its premises. Anyone who accepts a conclusion without being able to identify the premises from which the conclusion is drawn is not reasoning. Using Aristotle's distinction, knowing the conclusion amounts to an answer to a What-question, and knowing the premises amounts to an answer to the corresponding Why-question.

In short, if an investigator interprets someone's response as a conclusion due to human reasoning, there has to be an answer to the following question: "Why is that a conclusion? You have just made an inference, leading to this conclusion. So why did you draw this conclusion, rather than some other conclusion, or even no conclusion at all?" But the sole way to answer this question is by reference to the premises from which the conclusion is drawn. If there aren't any, the response can't be a conclusion due to human reasoning.

Three challenges could arise. First, it could be argued that this requirement is too strong since such reasons may be implicit without being explicit, for example in Karmiloff-Smith's (1994) model of representational redescription. True: a capacity may be present without being displayed. But what is equally true is that a non-display may be due to the absence of the capacity or to its presence in an attentuated form

(Flavell *et al.*, 1993, p. 328). The demarcation question between *not in competence* and *first in competence* is delicate, and is certainly not settled by reference to chronological age (Smith, 1993, p. 103). Second, it could be argued that a task may be designed such that the sole way to make a correct response is on the basis of the right reasons. In that case, expressing this reason is not required. Of course this could be so. But there are two qualifications. One is theoretical: there is a concession of the main argument that the right reasons are available to that individual. Indeed, without these reasons, the response would not count as "right" anyway (Chapman, 1988). The other qualification is empirical: it is also notoriously hard to ensure that this design feature is actually met. The research record shows that reasoning tasks have long been a battleground in this regard (Smith, 1992). Third, it could be argued that a good statistical analysis in terms of reliable differences makes the requirement for an analysis of reason superfluous. This challenge is flawed. It amounts to the conflation of reliability with validity. Statistically significant findings do not determine psychological validity.

5.2.3 *Foundational and Causal Accounts of Knowledge*

There are two standard accounts of knowledge, namely the foundational and causal accounts. These accounts lay down different conditions. They can be clarified using an example, such as knowing on Tuesday that today is Tuesday, i.e. knowing *p*. Under both accounts, knowing *p* requires one criterion to be met:

(a)   *p* is true.

This is because it is impossible to know a falsehood, though it is quite possible to believe a falsehood (Moser, 1995). Where the accounts differ is with regard to further conditions. Under the foundational account, there are two further conditions along with condition (a):

(b)   the knower believes *p*

(c)   *p* has a justification available to the knower.

Under the causal account, there are two other conditions along with (a):

(d)   *p* has a reliable, causal generation in the knower's mind

(e)   no other causal process is responsible for *p*'s generation in the knower's mind.

These are distinct accounts since conditions (a), (b) and (c) are different from conditions (a), (d) and (e). Both accounts continue to have sponsors (Goldman, 1986; Kenny, 1995; Moser, 1995; Nagel, 1997). It is evident that the capacity to justify is a strict requirement of the foundational account, but is not required by the causal account.

This means that the epistemological dispute as to which of these is the better account has important implications for developmental psychology.

There are several cases in which the capacity to give a reason (justification) is required for there to be knowledge at all. Modal knowledge is just such a case. Modal knowledge is the recognition of what "has to be" (Piaget, 1977/1986). There is a salient *décalage* in view of the fact that children can use *modal language* during the preschool years (Byrnes & Duff, 1989; Scholnick & Wing, 1995), but that *modal reasoning* is not typically present until late childhood (Miller, 1986), possibly adolescence (Morris & Sloutsky, 1998; Moshman, 1998), even adulthood (Bell & Johnson-Laird, 1998). Haack's (1978) definition of validity quoted in Chapter 4 is instructive here. A child who makes a correct response may do so because an argument *doesn't* have a combination of true premises and a false conclusion "in this context". This is a far cry from realizing that a valid argument requires that there *couldn't* be this combination "in any context at all". In short, a specific context may empower children to make a correct inference without a realization that the inference is necessary. Evidence about children's deductive reasoning requires their modal knowledge to be attested (Murray, 1981, 1990; Smith, 1993, 1999d,f). The practice implicit in most studies in developmental psychology has yet to match this requirement (Lourenço & Machado, 1996).

The foundational account always requires there to be a capacity to give reasons for a response since this is explicit in condition (c). The causal account does not make this requirement other than in cases where it is problematic as to whether conditions (d) and (e) are met. In consequence, the foundational account is the better account to use in the case of modal knowledge, such as the study reported in Chapter 6.

5.2.4 Internality and Externality of Relations

A distinction central to the revolution in philosophy at the turn of the last century concerned the internality and externality of relations. A classic argument accepted by Russell (1959) was that all relations have external terms. Take the example:

His left-hand glove is inside her right-hand coat pocket.

The relation *is inside* has two terms (*glove, coat pocket*). In this example, both terms have an independent existence since each can be characterized independently of this relation: the glove is his, whilst the coat pocket is hers. But not all relations have external terms. In a world consisting only of incongruent counterparts such as a pair of otherwise identical gloves, the left-hand glove is constituted by its relation to the right-hand glove and conversely (cf. Wittgenstein, 1972, 6.36111). The sole way to characterize either glove is by reference to its relation to the other. Bickhard (2002) has argued forcefully that the difference between the internality and externality of relations has been ignored in developmental psychology and that this oversight is disastrous. His argument generalizes to cover modal reasoning.

Modality is bound by internal relation since one modal concept, such as necessity, can be characterized only in terms of another (possibility). Thus a necessity is defined through, and so constituted by, negation and possibility, since a necessity is that which could not be otherwise (Smith, 1993, 1997, 1999d). Deductively valid reasoning is also bound by internal relations just because such reasoning is constituted by the relation between a conclusion and its premises (Johnson-Laird, 1999). Sound reasoning by mathematical induction requires the internality of relations to be respected just because the conclusion generalizes to any other number in the same system, i.e. the system in which the truths demonstrated in the premises were located. It follows that the assessment of knowledge in these cases requires a grasp of the internality of relation. The sole way in which this can be demonstrated is by recourse to the reasons for responses. It begs the question to suppose that children have this grasp at their disposal. Reliance on their correct responses is an inadequate ground for ensuring that this requirement is met.

5.2.5 *Functions of Statistical and Epistemological Analysis*

Each type of analysis has a distinctive function. This means that demonstration in an epistemological analysis is not generalization in a statistical analysis. So neither type of analysis can be used for the purposes of the other.

A statistical analysis is directed upon generalization. The analysis would run thus. The observed differences in this sample are such-and-such; this difference is statistically unlikely; therefore this same difference is true of the corresponding population. By contrast, epistemological analysis is directed upon the demonstration that a particular pattern of argument is identifiable in some case. The analysis would run thus. Reason such-and-such is displayed by children in this sample; this reason is an instance of type of reasoning so-and-so; therefore, reasoning so-and-so is within the capacity of children in this sample. Underlying such an analysis is the principle *ab esse ad posse*, i.e. whatever actually is the case can be the case. The demonstration is a verification that a possible type of reasoning is actually available in this sample. What requires identification is that a specified intellectual instrument (for example, reasoning so-and-so) "was actually at the subject's disposal. Here, whether we like it or not, is a question of fact" (Piaget, 1970/1977, p. 5; cf. 1998, p. 26).

A contingency table suitable for statistical analysis is not a set of reasons suitable for epistemological analysis. Even so, these may be combined in one study with a dual focus on both the causal conditions of an act and the rational conditions given for the judgment. A statistical model is a formal model. Equally, a logical model is a formal model. The former can be used in an analysis relevant to causal description and explanation. The latter can be used in an analysis of the reasons given for some judgment. Each can complement the other with neither reducible to the other. This distinction was not lost on Piaget (1953) who regarded logic as a formal model for use in science.

In short, a rationale has been set out for the use of a critical method in the collection of reasons for responses to be interpreted in an epistemological analysis. Modal knowledge has turned out to provide a good case in this regard in this rationale. This is significant in view of the contribution made by modality to valid deduction and to the necessitating inference in mathematical induction.

Chapter 6

Children's Reasoning by Mathematical Induction

In their study of reasoning by mathematical induction, Inhelder and Piaget (1963) — see section 3.3 — drew two conclusions, one that children aged 5–7 years could make this inference on the basis of iterated actions, and the other that their reasoning was modal. The present study set out to replicate and to adapt their study.

6.1 Method

Children. One hundred children took part in the study, drawn from Year 1 ($n = 50$) and Year 2 ($n = 50$) in three state schools in NW England covering the full range of normal ability during the first two years of compulsory schooling. There was an equal distribution of boys (mean age 71.9 and 83.8 months) and girls (mean age 71.0 and 83.6 months) in each school year within the range 64 to 93 months. One quarter ($n = 25$) of the children were aged five years, almost another quarter ($n = 23$) seven years, whilst the remaining children ($n = 52$) were aged six years. All of these children were interviewed twice in study I and again in study II. Each child was assigned an identifying tag, for example (I: 12) refers to child 12 in study I and (II: 94) refers to child 94 in study II. Children in school Year 1 were coded 31–60, 81–100; children in school Year 2 were coded 1–30, 61–80.

Design. The design was restricted in two ways. First, it was confined to middle childhood, based on children in the age range 5–7 years. Second, it was confined to reasoning about natural numbers in simple arithmetic. There were three hypotheses under which differences in performance related to school Year would be linked to:

(a)   children's reasoning by mathematical induction based on iterated action
(b)   children's modal reasoning based on logical necessity
(c)   children knowing how to count without knowing when to count.

This first hypothesis (a) was directed on repeatedly drawing a correct conclusion about "what is the case". As such, it is compatible with truth-functional logic, i.e. truth (correct responses) and falsity (incorrect responses) are at issue (Haack, 1978; Quine, 1972; Sainsbury, 1991). This hypothesis is compatible with Poincaré's position, if mathematical intuition is interpreted as iterated action. But this hypothesis is incompatible with the Frege–Russell position in which mathematical induction is reduced to logical deduction, if children's reasoning based on iterated action is not based on deductive logic (Inhelder & Piaget, 1963). Note that the Genevan account is frequently interpreted as a requirement for reasoning by logical deduction in number development (Bideaud, 1992; Case, 1999; Flavell *et al.*, 1993; Gelman & Williams, 1998; Nunes & Bryant, 1996; Siegler, 1996). But hypothesis (a) is the denial of this requirement.

Hypothesis (b) was directed on modal knowledge about "what has to be". Modal knowledge has long been central to Piaget's research programme (Smith, 1993, 1999d). And modal reasoning is central to recent research on the development of reasoning (Bell & Johnson-Laird, 1998; Morris & Sloutsky, 1998; Moshman, 1998). Children's mathematical reasoning could be linked with modal knowledge in either of two ways since all mathematical truths are necessities and all deductive inferences are necessities (Smith, 1997). Note that hypotheses (a) and (b) are psychologically independent since it is one thing to make a correct inference, and something else again to understand this as a necessity (Smith, 1993, 1997, 1998b, 1999d).

The third hypothesis (c) concerned the interaction of counting and reasoning. Research in developmental psychology has shown that successful counting is under development during the preschool years. Much less clear is how this is to be interpreted, notably in relation to numerical reasoning (Baroody, 1992; Dehaene, 1997; Gelman & Williams, 1998; Sophian, 1995). Further, arithmetic is central to the school learning of children aged 5–7 years with skills in numeracy at a premium (DfEE, 1997, 1999). Although counting abilities were not totally disregarded in Genevan studies, they were assigned a secondary place to the study of reasoning (Flavell, 1970; Gréco, 1963; Piaget & Garcia, 1987/1991; see also Bryant, 1995 and Nunes & Bryant, 1996, 1997). So there are two reasons why intellectual control over counting during early childhood should be of constructive value to children at the outset of schooling. Even so, there are two ways in which counting abilities can constrain children's reasoning. First, children might reason in contexts where counting is more appropriate, i.e. children run into error by substituting poor reasoning for correct counting. Second, children might count in contexts where reasoning is more appropriate, i.e. children run into error by incorrect counting instead of correct reasoning.

Procedure. Two cross-sectional studies were undertaken in which all the children participated. A subsample (*n* = 10) of these children also participated in an intervening microgenetic study, which is not reported here. Study I was undertaken in February–March and study II in May–June during the same academic year.

Five tasks were used in the two studies. Three tasks were common to both interviews, covering counting abilities (Counting), number reasoning (Conversation), and mathematical induction (Recurrence). One task was unique to study I covering task-instructions (Same). One task was unique to study II covering counter-factual reasoning about number (Teams). Four tasks are now reviewed in order of their use with the main (Recurrence) task reviewed last.

6.1.1   Same Task

- *Question 1: Here are three lines [pointing in turn]. Which two lines are the same?*
- *Question 2: Is there the same in each, or is there more in one than in the other?*

This task was confined to study I where it was presented as the first task. Its demands were compatible with level 1 criteria in National Curriculum Mathematics (DES, 1991; DfEE, 1999), namely the ability to count to ten. The purpose of this task was to familiarize the children with the typical form of the questions in the tasks ahead and also to allow them to be successful at the outset of the interview. Ten large red counters (each 4 cm in diameter) were presented in three parallel lines which were about 5 cm apart. Two lines had three counters with the remaining four counters in the third line aligned to the left so that its fourth counter protruded to the right. Each child was asked the two questions in the same order. A line with three counters was removed after Q1 had been answered. The order of the disjuncts in Q2 was randomized across the sample.

6.1.2 Counting Task

- *Question 1: Is there more in one line than in the other, or is there the same in each?*
- *Question 2: Is there more in one line than in the other, or is there the same in each?*

This task was presented either second or fourth in interview I, and either first or third in interview II. Order of presentation was counterbalanced for each child, and so this task was presented either just before or just after the Recurrence task. The task was adapted from Sophian's (1995) work. A different version of the task was used in each study in which a line of counters was attached to an elastic ribbon and then lengthened. The same question was asked twice, first before (Q1) and then after (Q2) the transformation. Children who answered Q1 without counting were invited to count the number of elements on each line as a check on their ability. Plastic buttons which looked like teddy bears in two colours (yellow, blue) were used. These buttons were attached to two strips of material whose length was equal at the outset. Each yellow bear was 2.5 cm, and each blue bear 2.2 cm, in length. One strip of material was black linen (20 cm in length) with four yellow teddy bears in a 17 cm line. The other strip was black elastic with five blue teddy bears in line. When non-stretched, its total length was 20 cm with the five blue teddy bears in a 13 cm line. When stretched, the elastic strip extended to 36 cm with the five teddy bears in a 20 cm line. The strips were initially referred to as lines by the investigator, or as teams by many of the children. Q1 was initially asked with the elastic in its non-stretched state. After an answer had been given, the investigator then said "Look at this [pointing to the blue teddies] — it is elastic". A demonstration followed and the elastic was stretched by the investigator. Some children then spontaneously tested the ribbon with the yellow teddies, and they were given informational feedback that linen does not stretch. Q2 was then asked.

In study II, plastic buttons which looked like red or orange cats were attached to two strips of material. All the cats were identical in length. One strip was black linen 27 cm in length with 12 green cats completely covering the strip from end to end. The other strip was black elastic and was also 27 cm in total length with 11 orange cats in a 24 cm line in its non-stretched state. When stretched, the elastic strip extended to 48 cm with the cats in a 45 cm line. At the outset, the investigator identified the two lines (or teams), and then pointed out that the orange cats were on elastic. This was stretched out, leading to Q1. The elastic was then relaxed and Q2 was asked. In study I, the team with the greater number of animals was the shorter line when the elastic was non-stretched, and longer when it was stretched. In study II, the team with the greater number of animals was the longer line when the elastic was both stretched and non-stretched.

6.1.3 Conservation Task

- *Directive 1: Help yourself and take enough whites so that you can make a line just the same as the blue line.*
- *Question 2: Is there the same here [pointing to the blue line] and there [pointing to the white line] or is there more in one than in the other?*
- *Question 3: Does there have to be the same in each, or not?*

This task was presented either second or fourth in interview I, and either first or third in interview II. Order of presentation was counterbalanced for each child, and so this task was presented either just before or just after the Recurrence task. The task was adapted from Piaget and Szeminska (1941/1952, Ch. 3). This was interpreted in terms of three parts, directive, truth-functional question, and modal question (Smith, 1993, pp. 63, 88). The investigator laid out a line of six blue counters (each 2 cm in diameter) in a 30 cm straight line on the table directly in front of the child. A transparent container with 25 white counters (each 2 cm in diameter) was adjacent to this line. The directive was then made, leading to the child's initial selection from the pile of 25 white counters. Note that Q1 is really a permissive invitation or implied command, not a question. So this study is a one-question conservation task (Porpodas, 1987; Rose & Blank, 1974). Children who selected more or less than six white counters were given feedback ("There are six in the blue line, but you have picked more/less than this"). All children eventually selected six white counters. The investigator then lengthened the line of six white counters such that the four middle white counters were approximately the same overall length as the blue line with one white counter protruding at each end (see Figure 6.1). All children were asked Q2 (correct reasoning) with the order of the disjuncts randomized across the sample. Subsequent to this, children were in all cases asked to give their reasons ("Why is that?") for their answer to Q2. At issue in Q2 was whether the children would make responses which were correct (true) rather than incorrect (false). Thus Q2 can be characterized as a truth-functional ("what is the case") question in contrast to Q3 which was a modal ("what has to be the case") question. In study I, the decision to ask Q3 was made on the spot. This decision was

After Q1   Before Q2

Note
⊗ = blue counter
⊕ = white counter

Figure 6.1: Array after Q1 and before Q2 in the Conservation task.

based on the children's apparent willingness to share their thoughts with the investigator as opposed to those children who did not seem so disposed. Since each interview included four tasks and the principal aim was to establish the children's thinking on all four, caution prevailed. Pressing Q3 might have a deflationary effect on thought (Dillon, 1990; Donaldson, 1992; Siegal, 1997, 1999). This policy of cautious restraint in study I turned out to be misplaced and so was set aside in study II. Once again, Q3 was always followed by a request for a reason ("Why is that?").

6.1.4 Teams Task

- *Question 1: In this team [pointing to . . .] and that [pointing to . . .], is there more blue teddies, or more blue cats, or is there the same in each?*
- *Question 2: The Blue Teddies want to play in fancy dress. Next time they play, they are going to dress up as blue cats. If they — the "team" of blue teddies [pointing to the "team" of blue teddies] — do that and they dress up as blue cats, would they [pointing throughout at the "team" of blue teddies] have in their "team" more players as blue cats, or more as blue teddies, or would they have the same?*

This task was confined to study II and was presented as the final task. It dealt with identity (equality) in a counterfactual context in line with Frege's position (1884/1950, p. ii; 1892/1960, p. 56) reported in section 3.1. There were two phases during which three questions were asked. Plastic buttons which looked like animals (cats, teddy bears) were attached to four black linen strips (25 cm in length). All the teddy bears and cats were identical in length (2 cm) and they were attached such that the length of the lines was identical (19 cm). Each was referred to as a team, and the Blue Teddies team was always presented first. Note that membership of this team was on the basis of two properties (blue and teddy bear) or their negations. In the context of this study, this quartet of properties was: blue (P), orange (not-blue $\neg P$), teddy bear (Q), cat (not-teddy bear $\neg Q$). So the four teams were:

Blue Teddies	$P \& Q$
Blue Cats	$P \& \neg Q$
Orange Teddies	$\neg P \& Q$
Orange Cats	$\neg P \& \neg Q$

The Blue Teddies had five members, whilst each of the other teams had four members. All the teams were initially on display together such that the three teams with equivalent members were presented in parallel columns and the team of blue teddies placed as a row across their end points. There were two phases, ostensibly for counting (A) and for reasoning (B). In phase A, the investigator removed two teams from the total array which remained visible. This was always the Blue Teddies and one other team in the order: Blue Cats, Orange Teddies, Orange Cats. The two selected teams were placed side by side and Q1 was then asked. Children who made incorrect responses were requested to count, which was completed successfully in all cases. In phase B, two teams with equivalent members were turned over so that their membership was invisible. The investigator then placed the Blue Teddies and one other team side by side, and Q2 was then asked. This was repeated for the other two teams.

6.1.5 Recurrence Task

- *Question 1: Is there the same in each, or is there more in one than the other?*
- *Question 2: Does there have to be the same in each, or not?*
- *Question 3: If you add [a great number, any number] here [pointing to one pot] and the same number to that [pointing to the other pot], would there be the same in each, or would there be more in one than the other?*

The Recurrence task was adapted from Inhelder and Piaget's (1963) study. A large pile ($n = 40$) of orange cats and another large pile ($n = 40$) of green cats were available on the table, one on the child's left and the other on the child's right. Two transparent plastic containers — which the children referred to as "pots" — were placed next to each pile. The children were invited to take a green cat in one hand and an orange cat in the other, placing each in its adjacent pot at the same time. This physical action of concurrent placement turned out to be difficult for some children who made temporally consecutive placements. The investigator always gave informational feedback in this event with a reminder of the directive that this (i.e. placing cats in each container) should be done at the same time. Three questions were asked. Children were asked Q1 in view of the base property of mathematical induction with Q3 relevant to its recursive property in line with the definition in Chapter 2. The children were also invited to make a modal response to Q2 in line with the second main conclusion drawn by Inhelder and Piaget (1963). These questions were repeatedly asked in four phases:

Phase A: initial state ($n = 0/n < 1$). The two piles of cats and two transparent pots were presented. In study I, both pots were empty and Q1 was asked. In study II, one green cat was deliberately placed at the outset by the investigator into its adjacent pot and the child was told: "You see these pots. Well, I am putting a green cat into its pot". Q1 was then asked. At issue in Phase A was the children's knowledge of the numerical contents of the two containers at the outset. Since performance was at ceiling levels in both studies, these responses were discounted in the subsequent analyses.

Phase B: actual and observed additions ($n < 10$). The children were then invited to take a green cat in one hand and an orange cat in the other, placing each cat at the same time in its pot. Successive additions then followed. Q1 was asked after three to eight additions had been made. The timing of this question was counterbalanced across the sample. After responding to Q1, Q2 was asked selectively in study I and invariably in study II.

Phase C: actual and non-observed additions ($n < 20$). The piles remained open to observation but the pots with their contents were placed out of view. Two large, black plant pots were used to screen/hide the smaller, transparent containers. The children verified that both black containers were empty, whilst the investigator said: "We'll put these pots in here so that we don't have to look at them. But you can still reach them, so let's keep on adding". Q1 was then asked to check that the transformation had not changed the child's previous response to Q1 at the end of phase B. In all cases, the same response was made. The children were then invited to continue adding cats to the pots in the same way as before. After several subsequent additions (amounting to the range 13–18), each child was asked Q1 and also asked to justify their response. This was followed by Q2 along the lines of phase B.

Phase D: hypothetical additions ($n = 6$/great number/any number). At the end of phase C, no more actual additions were made. The investigator said: "Let's stop adding these cats [pointing to the piles], and instead let's pretend to add, let's just think about adding cats". Each child was invited to pretend to add more cats in the same way as before and with the pots still screened from view as follows:

- pretend to add one to each pot at the same time
- pretend to add two to each pot at the same time
- pretend to add six to each pot at the same time*
- pretend to add a thousand to each pot at the same time
- pretend to add a great number to each pot at the same time**
- pretend to add any number at all to one pot and the same number to the other***.

Q1 was asked after directive * followed by Q2, and these were regarded as a "warm-up". The penultimate directive ** and ultimate directive *** led to Q3, one dealing with *a great number* and the other about *any number*. The ambiguity of the former required detection by specifying any particular number, in contrast to the latter which required an inference about any number at all.

Each child was individually interviewed with a separate clip-on microphone for both child and the author who acted as investigator in all interviews. Due to mechanical problems with a newly purchased audio-recorder, not all recordings could be transcribed, leading to an incomplete transcript for three children in study I. This recorder was changed after the interview with child 35 with no further breakdown. Overall, each child's answers to 15 questions across the two studies generated a pool of 3000 answers. Thus the interviews were intensive in two respects since these young children were repeatedly questioned and repeatedly challenged. The extent of the questioning was under the children's control, and a disinclination to answer was regarded by the investigator as a signal to move on. Children who were disinclined to express their reasons were not encouraged to do otherwise. The incidence of non-responding is reported in the analysis. The resulting dialogues varied in length and quality. One fifth of the children (26% in study I; 20% in study II) displayed open delight by laughing or smiling at some point during their interview, usually at the crux of the questioning! Many of these delightful expressions amounted to a veritable "knowing" smile — "knowing" rather than knowing in view of the fact that the children sometimes falsely believed that it was the investigator who had made a false move. Several children volunteered to take part in the microgenetic study or expressed regret on learning that they had not been invited to take part. Taken together, this amounted to experiential evidence that many children were pleased to take part in the study. They welcomed the fact that a stranger ("from the university") was interested in talking with them; they were treated as VIPs in an individual interview; their views were important enough to warrant an audio-recording with a clip-on microphone; there was an element of novelty since the tasks were different from their classroom tasks from which the interview may well have been a welcome break. With hindsight, an interesting oversight was noted for the future as to the incidence, nature and contribution of emotion, motivation and personality on understanding (Brown, 2001; Demetriou & Kazi, 2001; Donaldson, 1992).

The complete transcript for each child in both studies was read and classified twice, first by the investigator and then by a colleague in an independent classification. The answer to each question had two parts, a response and a reason for that response. The responses for statistical analysis were classified into three categories (correct, incorrect, not known). Disagreement was evident in just over 1 percent of the cases (study I: 22; study II: 17), resulting in a revised and mutually agreed classification. This dataset was used twice, both in a non-parametric analysis and then independently in the logistic analysis reported in Chapter 9.

The reasons for these responses were classified for epistemological analysis. This classification was undertaken in four steps and was undertaken by the investigator and a colleague. The first three steps were undertaken collaboratively until complete agreement was reached. The fourth step is reported in full below. The first step was the demarcation of the reason or reasons from the response, leading to the identification of the actual reasoning displayed by each child. Some children explained their view clearly and concisely, whilst others were discursive and elaborate; some children gave no reason, whilst others offered several; some children repeated themselves, whilst others offered novel explanations. The second step was to condense the reasoning by removal of incidental expressions so that its core meaning was explicit. In all cases, this core was formulated using the child's own words which were never summarized. The third step was to credit each child with one good reason available in his or her reasoning, namely the best reason given by that child. Once again, this reason was always formulated using the child's own words without addition by the investigator. If no reason was given, the reasoning was classified as "not known".

The final step was to classify one good reason in one of five categories: not known; observational knowledge or empirical induction; counting or number knowledge; logical or deductive reasoning; modal reasoning which covered: (a) pseudo-modal reasoning; (b) modality of action; and (c) modality of thought. The distinction between pseudo- and sound reasoning was invoked by Piaget (1926/1929) and Vygotsky (1994), and some version of this distinction fits modal reasoning. The distinction between the modality of action and the modality of thought is well defined in terms of the difference between practical and theoretical inference (von Wright, 1983a, 1984). Modal reasoning was classified as (b) when a modal expression was used about an action. Modal reasoning was classified as (c) when a modal expression was in relation to the object (content) of thought.

In general, two simplifying assumptions are made in this analysis. One concerned demonstration (instantiation, identification). These demonstrations are not presented as generalizations. The principal aim is to give examples of the children's reasons such that these examples could serve in answering the question "What in fact were the reasons?" The other assumption concerned simplification in that each child's reasoning was assigned to one-and-only-one category. Although each reason presented has been edited in this way, it is verbatim reasoning, i.e. the reason presented is a reason actually given by the child.

The only reasons used in the epistemological analysis are explicitly reported. Thus no inferences should be drawn with regard to non-reported reasons, nor about the generalizability of the reasons reported here. The sole purpose of the epistemological analysis of reasons is demonstration (instantiation, identification). The primary task was to ascertain which reason this is, from the child's point of view and using that child's own expression of his or her thought. Doubtless psychological questions can arise about the typicality of these reasons, or about the extent to which they transfer, or about their causal origin. These are good questions, but they are beyond the scope of the present study directed instead upon the epistemological question "What is the normative status of the reasons which children in fact give for their responses on these tasks?" Notice that a change in task-design is likely to result in different responses, maybe different reasons as well. "Causal facts" have a long established place in empirical psychology. But that still leaves open questions about which reasons the children gave for their responses and their reasons for their responses. "Normative facts" are also facts and these should have an equally important place in empirical psychology and epistemology (Smith, 2002a,b,c).

The findings are reviewed in two parts, statistical (non-parametric) analysis of responses, and epistemological analysis of reasons for responses. A logistic analysis of these responses is presented in Chapter 9.

6.2 Results: Statistical Analysis of Responses

This analysis is a quantitative review of the children's responses from the five tasks using school Year as an independent variable.

6.2.1 Same Task

This task was completed successfully by all school Year 2 children in both studies, and by almost all school Year 1 children in study I ($n = 47$) and II ($n = 49$). Feedback was given in the four cases of error, which was evidently due to these children's belief that this task was more complicated than it was. The subsequent pattern of responses of these four children was comparable to that of their peers, and their responses were retained in the subsequent analyses.

6.2.2 Counting Task

Correct responses to both questions are shown in Table 6.1. Both responses were made correctly by a fifth of the children (Year 1 = 7; Year 2 = 15). A further three fifths made one incorrect response for Q1 over both studies with a roughly equal distribution by school Year. The comparable total for Q2 was 73, once again almost equally distributed. A repeated measures Wilcoxon analysis revealed that the difference in performance over the two questions in study I by the Year 2 children was not significant ($Z = 1.732, p < 0.083$). Using the same analysis, there was a significant difference in performance both by the Year 2 children in study II ($Z = 5.301, p < 0.000$) and by the Year 1 children in both study I ($Z = 3.000, p < 0.003$) and in study II ($Z = 5.568$, $p < 0.000$). A Mann-Whitney analysis revealed that the difference due to school Year was significant in study I for both Q1 ($U = 950, p < 0.002$) and for Q2 ($U = 1100$, $p < 0.012$), but that neither difference in study II was significant (Q1: $p < 0.098$; Q2: $p < 0.103$).

Table 6.1: Correct responses from the Counting task for both Q1 (elastic non-stretched) and Q2 (elastic stretched) in study I and study II.

Study	I		II	
Question	1	2	1	2
Year 1	35	44	8	39
Year 2	47	50	15	45

6.2.3 Conservation Task

Correct responses to the three questions are shown in Table 6.2. The majority of the children made correct responses to Q1 in both studies. One tenth of the children made one incorrect response either in study I (Year 1 = 4; Year 2 = 2) or in study II (Year 1 = 3; Year 2 = 2). A Mann-Whitney analysis of the pooled responses to Q1 revealed

Table 6.2: Correct responses from the Conservation task for Q1 (initial selection), Q2 (correct reasoning) and Q3 (modal reasoning) in study I and study II.

Study	I	II	I	II	I	II
Question		1		2		3
Year 1	47	46	28	38	10[a]	13[b]
Year 2	48	48	39	43	13[a]	22[c]

[a] 16 missing cases
[b] 12 missing cases
[c] 7 missing cases

no significant difference by school Year ($p < 0.340$). But there was a significant difference in responses to Q2 by school Year ($U = 938$, $p < 0.013$). Further, a Wilcoxon repeated measures analysis of the response patterns to Q1 and Q2 was also significant ($Z = 4808$, $p < 0.000$). Maximal scores to this question across both studies were made by 61 children (Year 1 = 24; Year 2 = 37), whilst a further 26 children made one correct response (Year 1 = 18; Year 2 = 8). Both responses of 13 children were incorrect (Year 1 = 8; Year 2 = 5). Not all children were asked Q3 for the reasons given in the Procedure. Overall, two thirds ($n = 67$) of the children were asked this question in study I, and four fifths ($n = 81$) in study II. A Wilcoxon repeated measures analysis of these distributions did not reveal a significant difference in response patterns across study I and study II ($p < 0.127$). Maximal responses to Q3 were given by 11 children (Year 1 = 4; Year 2 = 7), and one correct response by 21 children (Year 1 = 9; Year 2 = 12). Both responses of 27 children were incorrect (Year 1 = 15; Year 2 = 12). A Mann-Whitney analysis of the pooled responses to Q3 did not reveal a significant difference by school Year ($p < 0.273$), nor were the separate analyses for study I and study II significant. A Wilcoxon repeated measures analysis of the overall response distributions to Q2 and Q3 did reveal a significant difference ($Z = 5.772$, $p < 0.000$). A Friedman repeated measures analysis of the overall differences in response to the three questions on this task was also significant ($x^2 = 69.516$, df. (2), $p < 0.000$)

6.2.4 Teams Task

Correct responses with overall mean responses to the three versions of each question on the Teams task are shown in Table 6.3. These questions generated ceiling and floor effects to Q1 and Q2 respectively and the outcome of a repeated measures Wilcoxon analysis indicated that this difference was significant ($Z = 7.611$, $p < 0.000$). Maximal responses were made by 74 children (Year 1 = 34; Year 2 = 40) for Q1 and by 35 children (Year 1 = 11; Year 2 = 14) for Q2. This increased to 98 for Q2 and 33 for Q2 using one incorrect response as the standard. Children in neither school Year made completely incorrect responses to Q1, but slightly more of the younger children made completely incorrect responses to Q2 (Year 1 = 29; Year 2 = 27). A Mann-Whitney

Table 6.3: Correct responses from the Teams task to three versions of Q1 (actual comparions) and of Q2 (hypothetical comparison) in study II.

Question	1			2		
	a	b	c	a	b	c
Year 1	38	48	47	13	16	17
Year 2	44	47	48	17	22	17

Key
Hypothetical comparisons between:

	a	b	c
	Blue Cats	Orange Teddies	Orange Cats
Blue Teddies			
P & Q	P & $\neg Q$	$\neg P$ & Q	$\neg P$ & $\neg Q$

analysis revealed a significant difference in response by school Year to neither Q1 ($p < 0.184$) nor Q2 ($p < 0.529$).

6.2.5 Recurrence Task

The review is in three parts, dealing with findings relevant to the base criterion of reasoning by mathematical induction, its recursive criterion, and finally modal reasoning.

Base criterion. At the outset, the children were asked whether or not the number in each container was the same. Correct responses were given by almost all children with regard to the initial equality in study I (91%) and the initial inequality in study II (99%). No child made two incorrect responses. The single incorrect response in study II was probably due to wandering attention. The incorrect responses in study I were apparently due to a perceived ambiguity. These children did not regard zero as a number (both containers were empty) and so answered under their belief that Q1 referred to all of the plastic cats on the table. So these children guessed the number in each pile on the basis of the apparent size of each pile. Feedback was given in that the number of cats in the pots was the intended reference and so the cats on the table should be disregarded. All children in both studies answered Q1 correctly after the first and second additions of one cat to each container. The children's responses about equality–inequality were scored after the following additions had been made: phase B: actual observed ($n < 10$); phase C: actual unobserved ($n < 20$); phase D: hypothetical ($n = 6$) addition. These are summarized in Table 6.4. Maximal scores, i.e. all six correct responses, were gained by most of the Year 2 children (85%: $n = 41$) but by less than half of the Year 1 children (41%: $n = 20$). The difference was reduced for a combined score of only one incorrect response in Year 2 (92%: $n = 44$) and in Year 1 (80%: $n = 39$). Correct responses are shown in relation to school Year for the six questions in the two studies, and this distribution was found to be statistically significant (Mann-Whitney $U = 682.00$, $p < 0.000$).

Table 6.4: Q1 correct responses from the Recurrence task during phase B (actual observed addition), phase C (actual unobserved addition), and phase D (hypothetical addition) by school Year in study I and study II.

Phase	B		C		D	
Study	I	II	I	II	I	II
Year 1	47[a]	46	48	40	49	25
Year 2	48[a]	47	48[b]	46	47[b]	44

[a] 1 missing case
[b] 2 missing cases

Recursive criterion. Relevant findings came from two final questions of the Recurrence task, one about the addition of *a great number*; the other about the addition of *any number*. Maximal scores, i.e. all four correct were gained by a quarter of the Year 2 children ($n = 13$) but by only one tenth ($n = 5$) of the Year 1 children. This difference was reduced for a combined score of only one incorrect response in Year 2 ($n = 26$) and Year 1 ($n = 17$) children. Correct responses are shown in relation to school Year for the four questions in the two studies in Table 6.5, and this distribution was found to be statistically significant (Mann-Whitney $U = 701.50$, $p < 0.000$).

Table 6.5: Q3 correct responses from the Recurrence task about the addition of a *great number* and *any number* by school Year in study I and study II.

Study	Great number		Any number	
	I	II	I	II
Year 1	21	14	35	24
Year 2	25[a]	19	46[a]	44

[a] 2 missing cases

Overall, this evidence indicates that the majority of the children completed this task successfully. The overall rate of successful responses for Q1 was 90 percent with most of the Year 2 responses (95%) correct. The overall rate of successful responses for Q3 was just under 60 percent with two thirds (68%) of the Year 2 responses correct. A Mann-Whitney analysis indicated that this overall difference in relation to school Year was statistically significant ($U = 40961.50$, $p < 0.000$).

Modal criterion. The frequencies in Table 6.6 show the incidence of modal reasoning on the basis of Q2 of the Recurrence task. Maximal scores, i.e. all six modally correct. were gained by only a small number of children equally distributed

Table 6.6: Q2 modal responses from the Recurrence task during phase B (actual observed addition), phase C (actual unobserved addition), and phase D (hypothetical addition) by school Year in study I and study II.

Phase	B		C		D	
Study	I	II	I	II	I	II
Year 1	18[a]	19	13	22	19	33
Year 2	17[a]	24	14[b]	26	28[b]	25

[a] 1 missing case
[b] 2 missing cases

in both Year 1 ($n = 3$) and Year 2 ($n = 4$). This rate doubled for a combined score of only one incorrect response in Year 1 ($n = 6$) and in Year 2 ($n = 7$). Correct modal responses are shown in relation to school Year for the six questions in the two studies. There was no indication of a significant difference by school Year in a Mann-Whitney analysis of this distribution ($p < 0.682$). Even so, the overall incidence of correct modal responses was encouraging, amounting to just less than one half (46%) overall with the correct modal responses of the Year 2 children at 48 percent.

6.3 Results: Epistemological Analysis of Reasons for Responses

At issue are the reasons used by the children for their responses. The analysis covers three tasks (conservation, teams, recurrence).

The analysis of these reasons is presented using the five categories, unless a novel line of reasoning was apparent. The five categories are:

- no reason
- empirical
- numerical
- deductive
- modal.

Recall that two simplifying assumptions were made. One concerned demonstration (instantiation, identification). These demonstrations are not presented as generalizations. The principal aim is to give examples of the children's reasons such that these examples could serve in answering the question "What in fact were the reasons?" The other assumption concerned simplification in that each child's reasoning was assigned to one-and-only-one category. Although each reason presented has been edited in this way, it is verbatim reasoning, i.e. the reason presented is a reason actually given by the child.

6.3.1 Conservation Task

This task had three parts comprising one request and two questions. The request was for each child to select enough white plastic cats from a pile to match a line of six blue plastic cats on the table. The truth-functional question concerned the equality of the lines after the transformation of the white line. The modal question concerned the necessity of the equality.

Directive Q1: initial selection   Pooling the responses in study I and II, almost all ($n = 89$) children made both selections correctly and no child made both selections incorrectly. Eleven children (I: $n = 5$; II: $n = 6$) made one incorrect response. Most of these children ($n = 9$) selected more than six white cats. These children typically focused on the pile of white cats which were highly attractive to two children who selected as large a number as they could grasp in their hands so as to play with them, 16 cats by one boy (II: 32) and ten cats by one girl (I: 37). It was quite clear from elsewhere in the study that both children could count proficiently. Six children were simply incautious, by taking a big handful in excess of six on the basis of observation. They realized their error after being invited to check (II: 41, 48, 72, 78; I: 84; II: 100). One child selected nine whites, placing them on the table so that the end points of both lines matched. After being invited to count the blue line, she declared that there were seven blues (I: 50). In another group was one child who selected six white cats, making the white line longer than the blue line (I: 68). When asked whether there was the same in each, he declared that there were more whites than blues. In a final group was one child who selected less than six white cats, making the white line shorter than the blue line. When asked whether there was the same in each, he declared that there was (I: 74). All of these children were subsequently invited to make two equal lines and to check this by counting. Q2 was answered correctly by five, and incorrectly by six, of these children.

Question 2: truth-functional reasoning   Across the two studies, about three quarters of the responses were correct and about one quarter incorrect (I: $n = 33$; II: $n = 19$) with 13 children making both responses incorrectly.

Conservation task — reasons for incorrect response to Q2   These responses fell into three groups:

- BLUES AND WHITES ARE NOT EQUAL
- MORE BLUES THAN WHITES
- MORE WHITES THAN BLUES.

Reasoning for the latter pair of responses was spectacularly distinctive.

"BLUES AND WHITES NOT EQUAL" (I: $n = 7$; II: $n = 1$)   This reasoning was well known. Two children miscounted, i.e. after making an incorrect response which they set out to explain by counting the white and blue cats incorrectly. One girl, who had initially selected nine whites for Q1, persisted in miscounting as nine both the six whites and the six blues (I: 50). One boy correctly counted the six whites but then miscounted the six blues as five (I: 87). In five cases, the children explained the non-equality in terms of the increased length of the white line. No reason was given in one case.

"MORE BLUES THAN WHITES" (I: $n = 7$; II: $n = 3$)   The reasoning for this response was distinctive. With the exception of two cases where no reason was given, this incorrect response was based on a common line of argument which was not immediately apparent.

In three cases, the reasoning seemed to be familiar in research on conservation in that the children's argument was based on the increased length of the white line:

● *more blues because you moved [whites] one on there and on there* (I: 66)
● *more in the blues because the whites have been stretched* (I: 98)
● *when they were like that, they were just the same, but you spreaded them [whites] out* (II: 72).

But this can't be right! This can't be familiar reasoning since these children had argued that the blue line had more than the white line. How could this be in view of the fact that it was the white line — not the blue line — which had been spatially extended?

A clue was provided by the children's own reasoning. These children believed that, after the transformation, the lengthened white line was spatially longer with fewer members than the blue line. From their perspective, the transformed white line was no longer the same as the original white line. It was not the same in that two counters had been disqualified from membership of "the" white line in virtue of the transformation (see Figure 6.1). From this perspective, a cat at the end of the transformed (white) line was no more a member of the "same" line than the "other" cats in the pile of white cats also on the table. Using this criterion, the transformed white line had four cats, whilst the blue line still had six. Therefore, they argued, there were more blue than white cats. Such reasoning is explicit in the arguments of three children:

● *more blues because that one's got six, and this has five, no four* (I: 49)
● *because there is 1, 2, 3, 4, 5 and [pointing to whites] there's four here* (I: 58)
● *more blues because there's only four [whites] left* (I: 62).

What is interesting in these cases is not miscounting but misconception. From their perspective, the white line after transformation had four cats which these children correctly counted! The "other" two white cats were not counted since they were no longer part of the white line. All is clear in two further justifications:

● *more blues because you've taken two away* (II: 85)
● *less because these two aren't there* (I: 96).

Quite simply, these children believed *you've taken two away* (II: 85). The white line is no longer the same since *these two aren't there* (I: 96). This is despite the fact that the two end whites were a couple of inches away and open to direct observation. Although this analogy was not used by these children, just as players in a football game may be sent off as a result of which the team is reduced, so the "team" of white cats had lost two members. Therefore: there were more blues than whites.

This is a spectacular argument for the non-conservation of number. Central to this argument is the defining criterion which lays down the properties which must be met by all of its instances. If a puppy is defined as an individual who is (a) young and (b) canine, then all infant dogs are (and must be) puppies, and no old cats are (nor can be) puppies. Similarly, if a line in an array is defined as something (a) equal (b) in length to any other line in that array, then the blue and white lines should be equal in length. Quite simply, these children invoked in their mathematical reasoning the principle that two lines should equal, i.e. their properties should be self-identical. This is comparable to the principle invoked by children in moral reasoning, namely the "principle of . . . equal action; that is, everyone should get the same treatment under any circumstances" (Damon, 1977, p. 75). This principle is invalid in the moral domain (Turiel, 1983, pp. 158–159). It is equally invalid in the mathematical domain. The decisive intervention made by the children who used this argument is their disqualification of the protruding white cats which infringe their criterion of equality in line length. Since a defining criterion lays down essential properties which determine what is and what is not a line, such reasoning is modal since essential properties are necessary properties (Marcus, 1993). However, these children did not use modal language to express their belief, nor did their reasoning have an explicit modal force. Patently, this defining criterion is false: it is not a requirement that two lines in the same array should be equal. Unequal lines may be in the same array. Further, these children assigned more importance to spatial identity than to numerical equality, which amounts to privilegization of the former over the latter.

"MORE WHITES THAN BLUES" (I: $n = 19$; II: $n = 15$)   The converse response was displayed by other children. Some children invoked the same defining criterion, namely that lines in the same array should be equal:

- *more in the whites because the blues would need to be one there* (I: 25)
- *more whites because they are not in a proper line* (I: 81).

After the transformation, the whites are not in a *proper* line since two end whites extended beyond each end of the blue line (I: 81). It is for this reason that the blue line *needs* more (I: 25). By assigning priority to the transformed white line, these children contended that the blue line would need more cats or that the white line had gained an improper advantage. Crucially, the use of their criterion generates the non-conservation of number.

The majority of children, however, reasoned in a familiar way and regarded the transformed white line as longer than the blue line. Especially striking is these children's use of a premise which is central to the reasoning for a correct response. Right now, it is sufficient to notice that in two cases the children miscounted:

- *there's seven blues and six whites* (II: 16)
- *more white because there's seven of them, and seven of them* (I: 41).

In most of the other cases ($n = 26$), the children reasoned about the increased length of the white line:

- *that one's stretched out, and that one's not* (I: 17)
- *they are more spaced out* (II: 37)
- *they [the two end whites] have moved along* (I: 55)
- *because there is a big gap there* (I: 82)
- *because there's holes in each one* (I: 92)
- *you've stretched them* (II: 99).

According to these arguments, there were more whites because the white line had been stretched, or because the cats had been moved, or because the white line now had big holes in it. In one case (I: 51), an incorrect response was immediately corrected on request for an explanation. In three cases, no reason was given.

Conservation task — reasons for correct response to Q2   Three quarters of the children (I: $n = 67$; II: $n = 81$) made a correct response.

NO REASON   In 13 cases (I: $n = 8$; II: $n = 5$), no reason was given with no child doing so in both studies.

EMPIRICAL   Just over one third of the reasons were empirical (I: $n = 20$; II: $n = 35$):

- *you have just made bigger gaps* (II: 02)
- *you've only stretched out the white ones* (I: 29)
- *because there's more gaps in the whites* (II: 77)
- *because you've spread them out* (I: 97).

In these cases, the children are reasoning in terms of the action (*stretching out the whites* I: 29) responsible for the transformation of the white line, or the outcome of this action (*more gaps in the whites* II: 77).

NUMERICAL   About one sixth of the reasons (I: $n = 10$; II: $n = 11$) were numerical:

- *they're both six* (I: 14)
- *there's 3 and 3 there, 3 and 3 there* (II: 18)
- *I counted them again* (I: 64)
- *same in each. Why? You tricked me because you put 2, 4, 6 white and 2, 4, 6 blue. [With a big smile] The things are quite easy* (II: 88).

In these cases, the children are reliant on their ability to count which enables them not only to quantify both lines correctly (I: 14), sometimes with proficiency (II: 18), even with metacognitive awareness of this ability (I: 64), and intellectual conviction (II: 88).

DEDUCTIVE   Most of the reasons were deductive (I: *n* = 28; II: *n* = 28):

- *you're not adding or taking away, you're just spacing* (II: 08)
- *it looks like more, but it is the same* (I: 32)
- *because it was the same number and it's still the same number* (II: 46)
- *because it doesn't change if you just stretch things out. It changes the length, it doesn't change one more to be there* (I: 47)
- *because if there is the same at the start, and you don't put any more on or take any away, it will be the same number all the time* (II: 56)
- *cos I didn't get any more out of there, and you didn't* (I: 60)
- *still the same number* (I: 61)
- *because you've spreaded them out, it still makes the same* (I: 65)
- *you've still got the same amount* (II: 67)
- *because I knew at the start of it and they're just farther away to just make it look like they're more* (I: 73)
- *because you stretched out, and you've counted them, and you know which ones they are* (I: 91).

In some cases, the reasoning behind a correct response was minimal with a bare reference to a shared property of both lines (I: 61; II: 67). In other cases, this same conclusion was elaborated in that *it was the same number* before and so *it still is the same now* (II: 46). Such reasoning was sometimes elaborated by the specific denial of factors cited by the children who made incorrect responses. Thus spacing out (II: 08), stretching (I: 47) and spreading (I: 65) are reckoned not to increase numerical quantity. In some cases, an elegant distinction between appearance and reality was invoked (I: 32), in other cases the same distinction was explicitly set out (I: 73). In two cases, the children offered the decisive argument: there is an initial equality; a transformation had neutral effects; so the equality is preserved (II: 56; I: 91).

MODAL   Modal reasoning in response to a non-modal question was displayed in three cases (I: *n* = 1; II: *n* = 2):

- *the same number. Why? Because them ones are not out* (II: 43)
- *you can put them downward, if you want* (I: 74)
- *because when they were together they were the same and they must be when they are apart like that* (II: 94).

One reason (I: 74) is modal, identifying another possible way in which the transformation could be made without altering the number in the white line. Another (II: 43) is a denial of the pseudo-modal reasoning reviewed above for the incorrect response "more blues than whites". This boy was denying that the equality in length of both was a relevant criterion since he specifically attested *them ones are not out*. Since they are not out, they should be included in the comparison which he correctly carried through. The third case (II: 94) is a realization that equality is a necessary relationship: if the number in the lines is equal when the lines are together, the number must be the same when the lines are apart.

Reverting back to the unsound reasoning for the incorrect response "more blues than whites", the key premise was that two lines in an array should be equal. In supporting their incorrect responses, children were interpreting this as a requirement for spatial equality. Deductive reasoning for a correct response was effectively a denial of this premise, manifest as the realization that numerical equality is compatible with spatial non-identity. What separates the reasoning behind correct and incorrect responses is a key premise (see Box 6.1). Premises (1)–(4) are common to both types of reasoning. On the basis of these premises, (5) is accepted either as an observed fact or as a inference from these premises. The key difference is due to the next premise, namely whether lines in an array are equal (identical). In the case of unsound reasoning, premise (6a) is interpreted as a requirement for spatial equality. But (6a) is contradicted by (5). Committed to (6a), these children get round this contradiction by accepting the numerical inequality in (7a). This is to jump from the frying pan into the fire on two counts. One is that (7a) contradicts premise (1), which these children had initially accepted. The other is that (7a) contradicts (6a) in as much as a numerical inequality is a non-identity. These children can evade the first contradiction by "bracketing off" their initial

---

Box 6.1: Sound and unsound reasoning in number conservation

Premises common to both sound and unsound reasoning

(1) The number in the blue line is equal to the number in the white line
(2) The blue and white lines are spatially identical
(3) The blue line is not spatially transformed
(4) The white line is spatially transformed
(5) The blue line and the white line are spatially non-identical

Unsound reasoning: conclusion drawn from premises (1)–(5)

   therefore

(6a) Lines in the same array should be the same in being spatially identical/equal

   therefore

(7a) The number in the blue line is not equal to the number in the white line

Sound reasoning: conclusion drawn from premises (1)-(5)

   therefore

(6b) It is not the case that lines in the same array should be the same in being spatially identical/equal

   therefore

(7b) The number in the blue line is equal to the number in the white line

---

acceptance of (1), and then evade the second contradiction by restricting the scope of (6a) to spatial identity. This may be a psychologically understandable response, but it is not epistemologically acceptable. By contrast, the reasoning for a correct response is different due to an acceptance (6b) which is the negation of (6a). These children then drew conclusion (7b), which is the same as premise (1). Their reasoning is both valid and sound, valid since non-contradictory and sound due to true premises and a true conclusion.

In general, it was noted in Chapter 3 that, for Frege, identity (equality) is the central notion in arithmetic and that Piaget and Szeminska had concurred. The fact remains that during intellectual development

> the principle of identity itself evolves in its meaning with the progress of thought [and of all the principles of logic] remains the least self-identical throughout its development. (Piaget & Voyat, 1968, p. 2)

The point is not that children's development is a progression from the absence of this notion to its presence. Rather, what is at issue is how this notion is formed and used by children during their intellectual development. No doubt it was for this reason that Piaget was disinclined to accept Bruner's account of conservation through identity on the grounds that possession of identity alone is not enough without its principled use (Smith, 1993, p. 90).

Question 3: modal reasoning   There were more incorrect (I: $n = 45$; II: $n = 46$) than correct (I: $n = 23$; II: $n = 35$) responses. The remaining cases were unknown (I: $n = 32$; II: $n = 19$). There was a notable asymmetry in the pattern of reasoning. In almost half of the cases, an incorrect modal response was made on the basis of modal reasoning, even though very few correct modal responses were made on the basis of modal reasoning.

Conservation task — reasons for incorrect responses to Q3   The reasoning displayed by these children was often distinctive. To make the same point: an incorrect response was all the same often combined with an interesting reason. Crucially, most of them were modal reasons.

NO REASON   An incorrect response was made without a reason in about one eighth of the cases (I: $n = 10$; II: $n = 17$).

EMPIRICAL   About one tenth of the other cases were divided between three categories. Some were empirical (I: $n = 5$; II: $n = 1$):

● *no, one is far away and one is not* (I: 80)
● *because if you push them back, it would be the same size* (I: 89)
● *not if you don't want it to* (I: 94)
● *because they are just spreaded out* (II: 97).

These are familiar reasons in response to the previous question without regard for the different demands made by the modal question.

NUMERICAL   Others were numerical (I: $n = 1$; II: $n = 2$):

- *no, if there was 10 there and 6 like that* (I: 53)
- *no, if you put them 2 there, and these 2 there, that makes 6* (II: 37).

These reasons refer to alternative ways of arranging the counters by specifying other numerical combinations.

DEDUCTIVE   Others again were deductive (I: $n = 4$; II: $n = 4$):

- *if you want to stretch one side and you don't stretch the other side, it wouldn't be the same* (I: 15)
- *if you got another white one and put it there, they wouldn't be the same* (II: 29)
- *because if you move one, then it's not the same number and then it wouldn't be the same* (I: 42)
- *if you put some more there, it isn't the same* (I: 64)
- *if you put some more in the whites, the whites will have more than the blues* (II: 64)
- *because if you added a few more, there would be more* (II: 68)
- *if you take one away off the whites, then the blues will have more* (II: 74)
- *if that one only had 5 in and that one only had 6 in, it wouldn't be the same* (I: 93).

Such reasoning has three properties. First, the reasoning is conditional with an explicit use of *if . . . then* in every case. Second, the reasoning was often an enthymeme in that a key premise was not explicitly formulated, no doubt because this was regarded by these children as too obvious to be worth stating on pragmatic grounds (Grice, 1989). Third, the reasoning was counterfactual. An alternative possibility was identified as a hypothesis with a consequent expressed sometimes in the indicative (*it isn't the same* I: 64; *the blues will have more* II: 74) but typically in the subjunctive (*there wouldn't be the same* I: 15; *there would be more* II: 68). To say the least, such reasoning was very interesting. It was testimony to the presence of rational capacities which were open to further development.

MODAL   Most of the reasons for incorrect responses were modal (I: $n = 25$; II: $n = 23$) and these fell into three categories:

(a) pseudo-modal

- *if you had a friend and are sharing sweets and you had more than your friend, it wouldn't be very fair* (II: 20)
- *well, it's not fair cos that's bigger and that's little* (II: 42)
- *if you think you need to put more then you put more, but if you think you don't need to put more then you don't. But if you think you want to put the same amount, then you put the same amount* (II: 52).

In the first two cases fairness was invoked, one based on a social analogy of sharing (II: 20) and the other on physical length (II: 42). In the other case (II: 52), needs (and also wants) were invoked as explanations of action. Although modal concepts are used in none of these cases, this reasoning amounts to human necessity, not to logical modality (von Wright, 1984).

(b) modality of action

- *you are not like changing them over like in math, you don't have to put the same amount in each* (II: 15)
- *well, you could take two away and it wouldn't be the same* (I: 19)
- *it doesn't have to be, no, you could do halves if you had a little thing that was sharp and split them in half* (I: 24)
- *you could put some more on* (I: 46)
- *because there's loads of different numbers you could get* (II: 63)
- *cos someone might pick them up and take them away* (I: 65)
- *could be or not, because you could swap them around* (I: 68)
- *if you put them back together and you take one away, it doesn't have to be* (I: 75)
- *you could rearrange them back again* (I: 81)
- *you might be doing some take aways in the number* (I: 85)
- *because if I was daydreaming, I could have put one in* (II: 86)
- *[helping himself to 3 more] you can just go like this and put them in a nice stack* (II: 88).

In some cases (I: 19; I: 46; I: 81) an explanation was given about a possibility (what could be the case) which is an alternative to an actuality (what is in fact the case), and in some of these cases the explanation was also analogical with a reference made to mathematics learning (II: 19; II: 63; I: 85) or to human activities (I: 68; II: 86). Indeed, one case (II: 88) was a marvellous demonstration of an alternative action which was performed for the benefit of the investigator who — this boy believed (no doubt he was a Vygotskian) — needed some assistance! Other cases amounted to a denial of a necessity, namely a denial that a specified action had to be carried out (II: 15; I: 24). Since a denial is usually ambiguous (if something is not-*X*, what is it?), it is interesting that a particular alternative is clearly specified in both cases, one based on mathematics learning and the other on human activity. Indeed one denial of a necessity was clarified by identifying an alternative (possible) action (I: 75).

(c) modality of thought

- *it doesn't have to be the same because it can be five* (II: 01)
- *it can be different* (I: 18)
- *because it could be different* (II: 40)
- *there doesn't have to be the same number* (II: 51)
- *you can have the most or I can have the less* (II: 54)
- *one could have more* (I: 72)

- *it doesn't have to be if you put more whites* (II: 84)
- *there might be if some child needs some, but there might not be if nobody wants any* (I: 87)
- *because you could have five in one and not six* (II: 90)
- *because there can be more on this side. If this one had twenty on, there would be more in the blues* (I: 97).

Some cases were a bare assertion of a possible alternative which was not further elaborated (I: 18; II: 40; I: 72). Other cases made reference to a particular alternative such as another number (II: 90), or to an alternative based on human activity (II; 54; I: 87). Other cases again were a denial of a necessity (II: 01; II: 51; II: 84) in which an alternative possibility was stated or implied. But all of these cases contained modal reasoning for an incorrect conclusion.

Conservation task — reasons for correct responses to Q3

NO REASON   No reason was given in a small number of cases (I: $n = 5$; II: $n = 1$).

EMPIRICAL   Equally small was the number of cases where the reason was empirical (I: $n = 3$; II: $n = 3$):

- *yes, so it matches* (II: 27)
- *because they were all lined up before* (I: 78)
- *because they are all together and then they are not and then they are bigger* (I: 82)
- *because you have spread them out and they can touch each other* (I: 100).

These reasons were unremarkable in their reference to observable features of the array or the actions responsible for them.

NUMERICAL   In a similar number of cases, the reasons were numerical (I: $n = 5$; II: $n = 6$):

- *because you haven't like added one on* (II: 06)
- *I put out exactly the same as the blues* (II: 22)
- *because if you put one more counter on, one will be 6 and one will be 7* (I: 26)
- *if you put another white in there, that would make 7, and that would make 6* (I: 28)
- *because there's six on one side and six on the other* (II: 36)
- *because there's 1, 2, 3, 4, 5, 6, and there's six there and there's six there* (II: 66)
- *yes, because six put to six* (I: 95).

These reasons are testimony to the children's ability to count correctly the number of counters in each line.

DEDUCTIVE   Other reasons were deductive (I: $n = 5$; II: $n = 18$)

- *because if you never had the same, it wouldn't equal the same on both sides* (II: 04)
- *because if you have that many and just stretch them out, it means you haven't got any more* (II: 09)
- *otherwise it wouldn't be equal* (II: 21)
- *because they were the same in the first place* (I: 22)
- *because all you're doing is just stretching them out and it has still got the same amount because you haven't added any more* (II: 28)
- *because you put as many out, it looks less but they are still the same* (I: 32)
- *because you've put them together and they're just a little bit wider* (I: 36)
- *because there's no more added or no less taken away* (II: 61)
- *because they've just been stretched out, so they're both the same* (I: 99).

Several reasons are standard justifications for a correct response to the truth-functional question (II: 09; II: 28; I: 36; II: 61; I: 99). Other reasons are enthymemes, but these could be expanded to ensure a valid argument, namely that nothing had changed the initial equality (I: 22; I: 32). This was made explicit in other reasons where the assertion of an equality is due to the elimination (negation) of anything productive of an inequality (II: 04; II: 21).

MODAL Finally, some reasons for correct responses were modal (I: $n - 5$; II: $n = 7$), and these divided into three categories:

(a) pseudo-modal

- *because it wouldn't be fair for that one if that had them up to there and that had them up to there* (I: 21)
- *if there's six, you need six more to make it the same* (I: 38).

These reasons make reference to fairness and needs and so result in human necessities, which amount to pseudo-necessities.

(b) modality of action

- *you can just make them wider or longer, or whatever you want, but don't add any more back in* (II: 02)
- *there has to be the same, so you can face them to each other like I did* (I: 08)
- *I can see three there and three there and that makes six* (I: 73)
- *there has to be the same in each because you have to take some away* (II: 87).

Some of these reasons identified alternative (possible) actions (II: 02; I: 08; I: 73). As such, they are genuinely modal. One identified a possible action compatible with the necessity (II: 02); another reason (I: 08) identified a consequence of an equality; the other reason (I: 73) was inconclusive since it refers to what can be seen when it should refer to the modal properties of what is seen. The final reason (II: 87) identified a possible action for removing the equality, rather than removing alternative possibilities to secure a necessity.

(c) modality of thought

- *there is the same . . . but they don't have to be the same, because if you added one more to the blue, there would be more than white* (II: 07)
- *it doesn't really make a difference because there's no more spaces, there's not any more in between, so there can't be more whites or blues* (II: 32)
- *because it's always got to be the same, because it's just stretched out so it's a longer length and it will still be the same if you stretched out the blue ones as well* (I: 47).
- *there's got to be the same amount* (II: 69)
- *it's got to be the same, because all it is stretched bigger* (II: 71).

One modal reason was inconsistent with the response for which it is offered as a reason (II: 07). An affirmative response was made to a modal question (*there is the same*), but this response was then contradicted in a modal denial (*they don't have to be the same*). In giving her reason, this girl was led to change her mind. In other cases, a necessity was explicitly asserted in a linguistic switch (*it's got to be* is used in answer to a question about *what has to be*) either by ruling out something (*it's a longer length*) which apparently could disqualify the necessity (I: 47; II: 71), or by specifying an invariant identity (*the same amount*) responsible for the necessity (II: 69). In one superb case (II: 32), the reasoning explicitly matched the definition of necessity (necessity is that which could not be otherwise) in as much as the equality of blues and whites is necessary because its negation is impossible (*there can't be more whites or blues*).

6.3.2 Teams Task

Although the two questions had parallel designs with a requirement for three comparisons to be made, their demands were different. Three numerical comparisons were required by the first questions and these were successfully made in more than 90 percent of the cases. By contrast, the second question required three inferential comparisons which were successfully made in merely one third of the cases (II: $n = 102$). Reasons are reported only for this inferential question.

Teams task — reasons for incorrect responses to Q2   Almost two thirds ($n = 198$) of the responses were incorrect.

NO REASON   In over one half of these cases ($n = 114$), the children gave no reason and so were content with a response such as

- *more blue teddies* (II: 84)

and declined to give an explanation as to why this would be so. This is a surprising finding. Since this task was always completed last, the children were in all cases asked to give their reason for their response. Further, the children had in general been remarkably willing to give their reasons for their responses throughout this interview. One interpretation of their reticence is in terms of their inability to make correct inferences, i.e. the children did not give a reason since they had no reason for their response. Such an interpretation is the attribution of a specific incapacity which could be explained in several different ways. One is that the children had no reason for their incorrect response which was not based on inference at all. The other is that the children did have reasons for a response due to inference but that they had difficulty in formulating these reasons, either because they did not understand them or because they did understand them but were beginning to realize that these reasons were wrong. Another is that since this task was presented last, the children did have reasons which they were disinclined to share with the investigator.

EMPIRICAL   None was given in this category.

NUMERICAL   In just under one third of these cases ($n = 55$), reference was made to the number in each team:

● *more in the blues because there's 5 there and 4 there* (II: 02)
● *if you counted them, there's 5, and if you counted them, there's only 4* (II: 19)
● *4 there and 5 there* (II: 99).

In these cases, the children were reliant on their ability to count the number of actual objects open to observation rather than the number of objects under the hypothesis. This is explicit in the first protocol (II: 19), and implied by the others. In over one tenth of these cases ($n = 18$), empirical knowledge based on the observational properties of the teams provided the reason for the children's responses:

● *because you are dressing blue cats into blue teddies* (II: 09)
● *if they swapped them round they would still have more* (II: 57)
● *if the blue teddies are blue and the cats are blue, they would be on the teddies side* (II: 92).

In these cases, the social action of dressing up into different clothes (uniforms, costumes, etc.) dominates the reasoning (II: 09), where this action (II: 57) or its outcome (II: 92) are empirical properties open to observational learning.

DEDUCTIVE   There was one case ($n = 1$) of unsound deductive reasoning:

● *if you change them round, they will still be the same* (II: 80).

This reasoning is unsound since its conclusion (more blue teddies) is incorrect. Yet the key premises invoked for this conclusion are both convincing and correct (identity is preserved through the transformation).

MODAL   In the remaining cases ($n = 10$), the reasons were modal as follows:

(a) pseudo-modality

In five cases, the reasoning was pseudo-modal, for example:

- *you need equal players in each team* (II: 01)
- *because that's their proper one* (II: 30)
- *if they say "I want to stay in my normal suit", these would win* (II: 38).

These reasons assumed that the blue teddies had a "normal" (II: 38) or "proper" (II: 30) suit of clothes, or even that there "needs" (II: 01) to be an equal number in different teams

(b) modality of action

- *because you are changing strips and you can't fit any more on that* (II: 05)
- *because you can see there's four there and five there* (II: 19)

The former (II: 05) was an affirmation of an impossibility about what can (cannot) be done in that the hypothetical change planned by the five blue teddies would require more than four actual strips. The latter (II: 19) was an instrumental modality directed on the numerical properties of the teams.

(c) modality of thought

- *because the teddies might not want to change into a cat* (II: 29)
- *because it could be different* (II: 40)
- *that's got to be more because there's five in that one* (II: 90).

The latter (II: 90) was a transposition of the modal *has to* used in the question to *got to* in this reasoning. The former (II: 29) was a valid possibility which is not relevant to the hypothesis under consideration. The possibility invoked in the other case (II: 40) indicated that this reasoning is directed upon actual objects (which could be different) when it should be directed upon the objects as constituted by the hypothesis (and so could not be different under that hypothesis).

Teams task — reasons for correct responses to Q2

NO REASON   Almost one third ($n = 31$) of the responses received no justification.

EMPIRICAL   In about one sixth of the cases ($n = 17$), the reasoning was empirical:

- *because they've changed their suits* (II: 04)
- *they would still be the same because they would dress up as cats* (II: 38)
- *the number would be the same because they've dressed up* (II: 63)
- *they would just be the same, you just put some clothes on them* (II: 91).

Such reasoning was directed on the change in dress (suits, clothes) open to observation. In a further sixth of the cases ($n = 16$), the reasoning was numerical:

- *same because five teddies you need five cat suits* (II: 01)
- *same, the [i.e. their] number is bigger than all of them* (II: 80)
- *same, four in that and five in that, they are only dressing up* (II: 97).

In such cases, the children used their ability to count the objects as specified under the hypothesis, disregarding the actual number of objects in front of them.

NUMERICAL   Sometimes, numerical reasoning was invalid with a correct conclusion drawn from an incorrect premise:

- *same, there will be the same number in their team because there's four in this and four in this* (II: 55).

In this reasoning, a correct conclusion has been drawn (the number is the same), but the reasoning was based on an incorrect premise (the original five blue teddies have been reduced to four after their change).

DEDUCTIVE   In almost one third of the cases ($n = 29$), the reasoning was deductive:

- *same amount, same number* (II: 03)
- *still have the same amount, they wouldn't have any less or any more* (II: 61)
- *they would be the same — you didn't take one away so they are as they are* (II: 71)
- *because we didn't add any more, so there is just the same amount* (II: 73)
- *they would just dress up and they won't change* (II: 79)
- *they're just changing costumes* (II: 80)
- *the number would the same — the only difference is that they're cats and they're teddies* (II: 89).

In some cases, this reasoning was based on identity: the number stays the same and so the amount is the same (II: 03), no doubt because the change in costumes is the sole difference (II: 89) and this is extrinsic, not intrinsic (II: 79; II: 80). In other cases, the reasoning was reversible in that there would not be any less or any more after the change (II: 61), since nothing more was added (II: 73), or taken away (II: 71).

MODAL   Modal reasoning was displayed in the following cases ($n = 9$):

(a) pseudo-modality

- *you would need to take one away and you would have exactly the same* (I: 01).

This case of pseudo-modality (what you need to do) is combined with an impeccable quantification about a consequential equality.

(b) modality of action

- *same because if there wasn't the same amount of suits, one of them would have to stay in the teddy bear costume* (II: 04)
- *if these played and one got cancelled, they would be the same, and that one would run away so he couldn't play again, so all these would be the same* (II: 42 — repeated twice more).

In these cases of instrumental modality, the reasoning was directed on what has to be done (II: 04), or what has not to be done (II: 42), by the teddies.

(c) modality of thought

- *there would still have to be the same number* (II: 56)
- *there must be more in the teddies than the cats* (II: 73)
- *the number might be the same because they have four and they have five* (II: 89)
- *might be the same because the blues might change* (II: 89).

In the last two cases, a (true) possibility was affirmed, namely that the number might be the same (II: 89). Such reasoning amounts to the dawning realization of this possibility. A realization of a corresponding necessity based on numerical identity (equality) was evident in one case (II: 56); in another (II: 73), a deductive consequence of this identity was explicitly drawn. Such reasoning matches the two main types of necessity, one based on equality and the other on entailment (Smith, 1999d).

6.3.3 Recurrence Task

The analysis is in three parts, covering the base and recursive criteria of mathematical induction as well as modal criterion. The reasons relevant to the base and recursive criteria came from three questions dealing with correct reasoning in the three phases of both studies. They were identified as B1 (observed addition), C1 (unobserved addition), and D1 (hypothetical addition). Modal reasons were gained in a similar way from corresponding questions in phases B2, C2 and D2. The six separate sets of reasons were combined in the classification, where absolute magnitudes refer to one set ($n = 100$) whilst percentages refer to the combined set ($n = 600$).

Base criterion. In study I, the success rate for correct responses was almost at ceiling level (B1: $n = 95$; C1: $n = 96$; D1: $n = 96$). The responses of two children were missing due to recording errors. In study II, the success rate for the initial response was initially at a similar level (B1: $n = 93$) which then declined (C1: $n = 86$; D1: $n = 69$).

Recurrence task — base criterion Q1: reasons for incorrect responses   Overall, the incidence of incorrect responding was small.

NO REASON   Most of the incorrect responses received no reason (4%)

EMPIRICAL   Some were justified empirically (3%):

● *more here and less there* (I: 55).

NUMERICAL   Others were justified numerically (2%):

● *because I've counted sixteen in there and not counted that one* (I: 64).

The first of these cases (I: 55) amounts to circular reasoning in that an incorrect response (there was not the same in both containers) was justified by a premise which re-states this same conclusion. The tendency to rely on counting was evident in one case (I: 64).

DEDUCTIVE   Finally, there were some deductive reasons (1%):

● *the same in each, because you told me to add one to each, then they weren't the same, then I kept on carrying on* (II: 25)
● *same in each because if you put 6 in there and 6 in there, there would be the same in each* (II: 89).

Both deductive arguments (II: 25; II: 89) were notable as valid but unsound reasoning, valid in virtue of the use of a Euclidean axiom (*equals added to equals are equal*), but unsound since both children had drawn a false conclusion.

MODAL   No modal reasons were given.

Recurrence task — base criterion Q1: reasons for correct responses

NO REASON   No reason was given in about one sixth of the cases (16%), and this was typically for D1 with almost twice as many absent reasons in study II ($n = 46$) as compared with study I ($n = 25$).

EMPIRICAL   One third of the reasons (34%) were empirical:

● *more in the green, there's a gap there* (II: 02)
● *because they had a head start* (II: 63)
● *because I am popping them in at the same time* (I: 87).

These children used observation or recall applied either to their actions or to the objects as the basis of their (correct) belief.

NUMERICAL   About one fifth (23%) of the reasons were numerical:

● *because that's got 9 and that's got 8* (II: 03)
● *[counting] more in the greens* (II: 15)
● *the same because there's 18* (I: 17).

These children were proficient at counting, basing their correct responses on their use of this ability.

DEDUCTIVE   About one eighth (13%) of the reasons were deductive:

- *we put one in there and there was none in here, and we have just added them one at a time, so this one is going to have one more and this one is going to have one less* (II: 01)
- *[if we do this] there will always be more in the greens* (II: 04)
- *because if there was the same in each and you added one more in each, it would be the same* (I: 06).

These cases are inferential and correct. A version of the Euclidean axiom is explicit in one case (I: 06) and implicit in another (II: 04). A logical deduction derived from prior addition is also evident (II: 01). Although these children may have realized that these deductions are necessities, modal reasoning was not explicit in their arguments.

MODAL   Modal reasons (4%) were displayed equally over the two studies:

(a) pseudo-modality

- *because it wasn't fair at the start* (II: 73).

There was one case of pseudo-modality, manifest as the belief that the two containers required *fair* treatment and so have equal contents at the outset (II: 73). This reasoning is modal since it concerns the deontic modality "what should be" rather the alethic modality "what has to be" (Marcus, 1993; von Wright, 1983a). But according to this reasoning, it is normal (standard) for two lines to be equal with respect to their properties, and so it is not fair for two lines to have unequal properties. This is of course not the case, and so amounts to pseudo-modal reasoning.

(b) modality of action

- *you can see this one has got that much* (II: 01)
- *I can't think* (II: 14)
- *because you could have added one on to the greens* (II: 24)
- *because all the space in there got covered up before that, you can see this little arrow there and not in there, so I can tell* (II: 42)
- *because there was the equal number or some people think it was equal, but it's not because they might not have seen you put one in* (II: 47)
- *because you can see the three and you can see the three, so you just count one more and then it makes seven* (I: 65)
- *you can't see them, but when you see them you can count* (II: 68)
- *because I could have put loads in them and just two in these, but I didn't* (I: 71)
- *because I can see there's two there, two there, and one there (orange), and just the same in each* (I: 94)
- *more in the greens, because you might like put two in at the same time* (II: 96).

These cases were classified as modality of action since they were instrumental necessities about "what has to be done", i.e. which actions were believed to be possible or necessary for a state of affairs to be the case. The reasoning of these children concerned the action(s) which *you might like to do* (II: 96), which *you could do* (I: 71), which *you can do* (I: 65), and which *you can't do* (II: 68). This line of reasoning was testimony to children's inventiveness.

(c) modality of thought

- *there might be 1 more in there because we started off with 1 more in there* (II: 07)
- *they look the same. Are they really the same? It could be really the same, 'cause there's the same there and the same there, but we don't know inside* (I: 08)
- *because the last time it might be 14 and 13* (II: 15)
- *because there's 4. I counted 2 and I counted 2 and then there's another 1 and there has to be 5, and then I did it the same there* (I: 20)
- *you started off with 1 in there and 0 in there, then you put 6 in that and 6 in that, so there should be 7 in that* (II: 22)
- *because you started off with 1 in the green and 0 in the orange, so if you put 2 in both, 1 in each at the same time, then it will have to be the same* (II: 55)
- *because it's always got to be the same if you started off with 1 and you didn't in the orange* (II: 56)
- *because I just counted 6. If I put one more in each colour, there must be 7* (I: 66)
- *2 in there and 1 in there, so there must be loads in there* (II: 69)
- *because the greens had a head start, and no one notices, then the greens must have more* (II: 88).

These cases split into two types. One type concerns beliefs about alternative possibilities, for example *there might be one more in there* (II: 07), *it could really be the same* (I: 08), or *it might be fourteen or thirteen* (II: 15). They amounted to an awareness of several possibilities coupled with a reluctance to draw an overhasty conclusion about which one is, or has to be, the case. The other type concerned necessities. Some children explained that there *should be seven in that* (II: 22) or that *there must be seven* (I: 66) or that *there has to be five* (I: 20) or that *it's always got to be the same* (II: 56). In other words, these ten cases were beliefs about necessities on the part of young children.

Recursive criterion. Reasons relevant to the generalizing criterion of mathematical induction came from two versions of question Q3, about adding a great number and any number.

Great number   Just over half ($n = 52$) of the responses in study I were incorrect, of which two thirds ($n = 35$) were unjustified. The remaining responses ($n = 46$) were correct of which one sixth ($n = 8$) were unjustified. The responses and so reasons of two children were missing. In study II, two thirds ($n = 67$) of the responses were incorrect. Almost all of these ($n = 64$) were unjustified. The remainder ($n = 33$) were correct responses, and all of these were justified. Thus overall, the incorrect responses were typically unjustified, whilst correct responses were typically justified.

Recurrence task — recursive criterion Q3 "great n": reasons for incorrect responses

NO REASON   There were no cases, i.e. some reason was given for all responses. In itself, this could be testimony to the children's recognition of the importance of this question.

EMPIRICAL   Empirical reasons ($n = 5$) for the incorrect responses included:

- *still more* (II: 02)
- *because we keep putting it in at the same time* (I: 14)
- *because there would be loads in there and loads in there* (I: 21).

These reasons refer to actions (I: 14), properties of objects (I; 21) and outcomes of these actions on objects (II: 02).

NUMERICAL   Numerical ($n = 8$) reasons included:

- *one more in there* (II: 04)
- *10, 10* (I: 32).

The same cardinal number is specified in one reason (I: 32), and invoked as the outcome of a numerical operation in another (II: 04).

DEDUCTIVE   Deductive reasons ($n = 5$) included:

- *there's the same in the orange and the same in the green because I kept adding one in each and it makes the same* (I: 13)
- *because they are the same numbers* (I: 99).

The former (I: 13) amounts to this child's version of the Euclidean axiom. The latter (I: 99) could be the same argument, though the use of the plural *numbers* rather than the singular *number* indicates that this reasoning is directed upon actual objects (green objects and orange objects added to the two containers) rather than an invariant abstract object (one and the same number). An analogous interpretation fits the former reason as well since the outcome is regarded as a causal consequence of prior action.

MODAL   Modal reasons ($n = 2$) for an incorrect response were:

- *it will have to be the same number* (II: 06)
- *you've put a great number in that and a great number in that, well it does have to be the same, doesn't it!* (I: 19).

Both cases concerned (c) modality of thought. The modal element in both reasons is explicit, concerning "what has to be". Even so, these reasons are invalid and unsound since they embody the fallacy of ambiguity. The reference of the expression a *great number* is not fixed in these reasons and so the sense in which this expression is understood is indeterminate. Nonetheless, these reasons are interesting in view of the fact that modality was never mentioned by the investigator when asking Q3. Evidently autonomy, here manifest as the spontaneous use of modal reasoning, and modality, namely the appropriate use of modal reasoning, are not the same thing.

Recurrence task — recursive criterion Q3 "great n": reasons for correct responses

NO REASON   No reasons ($n = 8$) were given in a minority of cases and these were all confined to study I.

EMPIRICAL   Empirical reasons ($n = 6$) included:

- *that one because it's orange* (I: 26)
- *if that one's higher than that, there would be more in there* (II: 48).

These reasons concerned the observable properties of actions or objects.

NUMERICAL   Numerical reasons ($n = 40$) included:

- *because if you add 20 and 25* (I: 02)
- *if you put 1 in that and 2 in that, it'd be the same, but it would be more* (I: 23)
- *one thousand and one million* (II: 62)
- *depends which great number you put in* (II: 70)
- *I don't know because you never said the number* (II: 93).

There were individual differences as to what counts as a great number. Some reasons made very modest assumptions about great numbers which could be as small as 1 in contrast to 2 (I: 23). Other reasons referred to stereotypical representatives of the decade system, such as a thousand or a million (II: 62). Several reasons contained an elegant generalization dependent upon whether or not the great number was the same (II: 70; II: 93).

DEDUCTIVE   Deductive reasons ($n = 10$) included:

- *if they were both the same, they'd be the same amount* (I: 27)
- *if you take all of them out, and put the same in, it will be the same* (II: 38)
- *if you add the same great number, you'll just get the same amount* (I: 68)
- *if you put not a great number and then a great number, it won't be the same* (I: 96).

It is instructive to notice that these reasons were linguistically expressed as *if . . . then* conditionals, which these children had spontaneously used in the organization of their mathematical reasoning. These children tended to focus on the ambiguity of *a great number* by specifying what this is or could be without regarding the previously unequal amounts in the two containers.

MODAL   Modal reasons ($n = 15$) for correct responses included the following four cases of the modality of thought, and one of these was spectacular (see Box 6.2):

- *it doesn't have to be the same because there's lots of great numbers* (I: 47)
- *that might have a gillion and that might have a million* (II: 54)
- *because that might be a million and that one might be a hundred* (I: 100).

The first case exploits the indeterminacy of the expression *great number* in as much as there are different great numbers. The modality of this belief is explicit since the great number *might be a million or a hundred* (I: 100). The second case is a novel creation, in this the postulation of a new great number which *might [be] a gillion* (II: 54). Indeed, there are in general *lots of great numbers* and so *it doesn't have to be the same* great number (I: 47). The logic of this modal inference is impeccable (Smith, 1999d). Suppose a proposition p is necessary ($\Diamond p$). This is definitionally equivalent to the impossibility of its negation ($\neg \Diamond \neg p$). Now the affirmation of this negation amounts to the possibility of this negation ($\Diamond \neg p$). But this entails that p is not necessary at all. One case (II: 32) was a spectacularly successful display by a seven-year-old child of mathematical reasoning which was autonomous, analogical, modal, valid and sound (see Box 6.2).

**Box 6.2: John's reasoning about adding *a great number* in the
"equals added to unequals" study**

Interviewer	*How about if you put a great number in that one and a great number in that one. Would there be the same in each or more there or more there?*
John*	That would be right up to the cover in the sky and that would be right up to God, so then they would still have to be more.
Interviewer	*What's the cover in the sky?*
John	It's on top of where God lives

This reasoning is based on a distinctive analogy whose initial comprehension defeated the interviewer. As such, this reasoning matched Piaget and Inhelder's (1961) argument for the use of a critical method in Chapter 5, since it defied initial interpretation by the investigator.

This reasoning also had five normative properties which are reviewed in Chapters 7 and 8 under an AEIOU framework:

- autonomous
It was a free mental act which could not in principle have specifiable conditions, even though it does have an individual characterization.

- equality/entailment (modal knowledge of necessity)
It concerned "what has to be" in that all mathematical truths are necessities. This is a distinctive modal realization by a child aged seven years.

- intersubjective
It was and in line with the Euclidean axiom *equals added to unequals are unequal* (Heath, 1956), which is a paradigm case of "common ground" between different thinkers.

- objective
It was justified as a response in a valid (truth-preserving) argument.

- universal
It had a degree of generality, whether or not open to transfer under different causal conditions. As such, it amounted to (a level of) knowledge of universality.

* not the child's actual name

Any number In study I, four fifths ($n = 81$) of the responses were correct, of which just over a third ($n - 28$) were unjustified. About one third ($n = 5$) of the incorrect responses ($n = 17$) were unjustified. In two cases, the responses were unknown. In study II, two-thirds ($n = 68$) of the responses were correct, of which one sixth ($n - 11$) were unjustified. All of the remaining responses ($n - 32$) were incorrect, of which half ($n - 16$) were unjustified.

Recurrence task — recursive criterion Q3 "any n": reasons for incorrect responses

NO REASON Some reason was given for all of these responses.

EMPIRICAL Empirical reasons ($n = 14$) included:

- *more in that and none in here because we would need to put some in* (I: 45)
- *more in the other, because one is higher and the other is lower* (I: 68)
- *same in each because I went on and on and on* (II: 99).

Recall that the content of the containers in study I was initially equal, and initially unequal in study II. In these reasons, this difference is linked to a focus on objects and their properties in study I, but on the action of addition in study II.

NUMERICAL Numerical reasons ($n = 6$) included:

- *there would be less because both cats wanted 10, but one cat thought he had 10 and the other cat thought he had 10 when he had 9* (I: 37)
- *same in each because we put a thousand in before* (II: 44).

These reasons made explicit use of a specified cardinal number. The basic elements of "folk psychology" underpin the attribution of mental states to (plastic) cats, manifest as the attribution of desires in one case (I: 37).

DEDUCTIVE   Deductive reasons (*n* = 4) included:

- *6 in each because if you put 6 in there and 6 in there, it would be the same* (II: 64)
- *same in each because if you put the same number in there and in there, it would be the same in each* (II: 89)
- *same in each because if you put the same in each, they will be these same in each, because there would be* (II: 100).

These reasons were intriguing since they all arise in the study where the containers had unequal contents. The first reason (II: 64) was a self-identical substitution-instance generated jointly by the question (any number) and the previous question about the hypothetical addition of six to each container. This child not only knew the quantity in each, but could also use this knowledge in a deductive argument. The second reason (II: 89) generalized this argument by means of a Euclidean premise. The third reason (II: 100) contained the same generalization, which is itself given a further reason (*because it would be*).

MODAL   Modal reasons (*n* = 4) sub-divided into three categories:

(a) pseudo-modality

- *more there because you have not put the right number in* (I: 91).

One case concerned pseudo-modality, shown by reasoning about the *right number* (I: 91), presumably because this child assumed that each container should have an equal number.

(b) modality of action

- *same in each because you could be putting the same amount in each* (II: 97).

One case concerned modality of action, namely the possibility of *putting the same amount in each* (II: 97).

(c) modality of thought

- *that might have sixty-six and that might have sixty-three — they're not the same number* (I: 21)
- *[taking a peep inside] if the orange is fuller than the green, that means there must be more in there* (II: 94).

Two cases concerned modality of thought, one about an alternative possibility (I: 21) in disregard of the constraint that the number — whatever it is — is the same for both. In the final case (II: 94), necessity was invoked in the inference from an unquantified volume observationally verified to a conclusion about an intensive magnitude (Smith, 1993, p. 166).

Recurrence task — recursive criterion Q3 "any n": reasons for correct responses

NO REASON   Some reason was given for all of these responses.

EMPIRICAL   Empirical reasons ($n = 48$) included:

- *more in the greens, there's always been more in the green* (II: 03)
- *because you put loads in there and loads in there* (I: 17)
- *because you see there the same in both tubs* (I: 29)
- *more there because that one had a head start* (II: 43)
- *same number in that and any number in that* (I: 55)
- *because you keep just putting one more in each, and it makes it just equal* (I: 75)
- *more there because you put one in and I put none in and keeped on doing it* (II: 69).

These observationally based reasons were varied, and included: centration on the "head start" (II: 43); empirical inductions from actions and objects (such as II: 03; II: 69; I: 75); distorted responses triggered by the question (I: 55); indeterminate claims (I: 17), and observational reports (I: 29).

NUMERICAL   Numerical reasons ($n = 27$) included:

- *if you added a million in there, there would be a million, and if you added a million in there, there would be a million* (I: 02)
- *definitely would be one more* (II: 02)
- *it would be just a million and 6 in there and a million and 6 in there* (I: 04)
- *100 in that, 100 in that* (I: 64)
- *100 there and 100 there and then that one would have loads more* (II: 66)
- *if there was 5 in there and 6 in there, they would be a little bit bigger* (II: 92)
- *because they are the same numbers* (I: 99).

Some reasons were numerical tautologies (I: 02; I: 04), whilst other reasons due to the same child (II: 02) were based on precise quantification. Other reasons combined precise quantification with numerical indifference (II: 66; II: 92). Some reasons left equality open, for example by invoking a number twice over (I: 64) or by referring to the same *numbers* — rather than *number* — in the two containers (I: 99).

DEDUCTIVE   Deductive reasons ($n = 27$) included:

- *if we started with putting 0 in there and 1 in there, and then adding 5 and 5, and so it wouldn't be the same — this would be 5 and 6* (II: 01)
- *because, say, you're adding 5 to that one and 5 to that one, they'd just be the same because you're adding the same number to each* (I: 24)
- *because you said any number to that and the same to that, so it is going to be the same, isn't it?* (I:32)
- *because if you put 6 cats and 7 cats in, it won't be the same, but if you put 7 cats in and 7 cats in, it would be the same* (I: 59)
- *that one had the head start, so that would have more because you put the same amount in* (II: 62)
- *same in each because you're putting in any number you like and it's actually the same number, because you're adding your favourite number into one and your favourite number into the other pot* (I: 88).

In some of these reasons *because* (I: 24; I: 32; I: 59; I: 88) rather than *if ... then* (I: 59; II: 01) was used to link premises to their conclusion, whilst *so* is used in others (I: 32; II: 62). Different Euclidean axioms (*equals added to equals are equal; equals added to unequals are unequal*) were appropriately adapted in some reasons, sometimes in an abbreviated form (I: 32) and sometimes elaborated using numerical premises (II: 01). There was also a delightfully egocentric interpretation of the investigator's *any number you like* which one child interpreted as *your favourite number* (I: 88).

MODAL   Modal reasons (*n* = 8) fell into two categories.

(b) modality of action

- *more there, because you might just leave one there* (II: 88)
- *I'm putting like a number in the orange pot and a number in the green pot, and it's got to be the same* (I: 69).

There were two cases of the modality of action. In one, the reason invoked a permissible action (II: 88). In the other, a necessary deduction was drawn from prior activity (I: 69).

(c) modality of thought

- *more there because there might have been more there [grinning]* (II: 05)
- *it might be even, actually it could be sometimes, it could be odd* (I: 08)
- *because they always put one in the green and none in the orange, so it must be always the green which has more* (II: 14)
- *there might be more in there* (II: 58)
- *last time [referring to Q3 about a great number], they could have had a higher number and the greens could have had a lower number* (II: 63)
- *I don't know [why], but if it's the same, it's got to be the same* (I: 73).

There were several cases of the modality of thought. One reason (II: 14) captured the modal nature of deductive validity in line with standard logic texts (Sainsbury, 1991). Another is distinctive (I: 73) in as much as an initial assertion of an incapacity is reanalysed in terms of an insight which is central to intensional logic, namely that all identities (equalities) are necessities (Kripke, 1980; Marcus, 1993). In several cases (I: 08; II: 05; II: 58), there is a realization that alternative possibilities may have been instantiated (II: 63).

Modal criterion. There were three modal questions in each study and each was asked only after its corresponding questions dealing with correct reasoning. Pooling the responses revealed that, in study I, just over one third (36%) of the responses were correct, half were incorrect (50%) and the remainder (14%) were unknown. The unknown reasons were due either to recording error (six cases out of 300) or to the investigator's decision — made on the spot — *not* to press for a reason for each-and-every modal response in study I (35 cases out of 300). The findings from study I showed that such restraint was not necessary. The children who were unwilling or unable to give a reason made this very clear themselves without requiring an adult's assistance. In consequence, this cautious policy was set aside in study II, where overall the breakdown of modal responding was almost even (44% correct; 56% incorrect).

Recurrence task — *modal criterion Q2: reasons for incorrect responses* Overall, more than one half (56%) of the responses were incorrect.

NO REASON   No reason was given in one fifth (20%) of the cases

EMPIRICAL   Empirical reasons (4%) included:

- *if it hadn't had a head start, it would be the same* (II: 80)
- *because sometimes you get it fuller* (I: 89)
- *because my sister and me both wanted two biscuits, but I wanted two and my Bethany wanted two, so my mummy gave us two* (I: 37).

One reason is testimony to children's capacity to reason analogically based on social relationships (I: 37). Other reasons focused on initial or eventual states (II: 80; I: 89).

NUMERICAL   Numerical reasons (3%) included:

- *there's 11 in there cos I counted this pot* (I: 48)
- *don't know because I haven't counted them* (II: 58)
- *because if you don't count them you won't know* (II: 74)
- *if you put three oranges in, and two green in* (I: 94).

These reasons revealed that counting was sometimes viewed as one way (I: 48; I: 94), and sometimes as the only way (II: 58; II: 74), to make a numerical response.

DEDUCTIVE   Deductive reasons (4%) included:

- *cos if we add these 2 to there, then it means that there is always more* (I: 08)
- *because if we tip them out and put in each one at the same time, we'll get the same* (II: 39)
- *because we started with 1 in there and 0 in there and we are pretending to put 6 in both, so it would be the same in both* (II: 57)
- *because if you put 7 in there and 2 in there, there won't be the same. But if you put them in the same amount in there, will be the same* (I: 90)
- *because if that one was the same as that, it would be a draw* (II: 92).

Although these children had made incorrect responses, they were nonetheless capable of reasoning deductively. Their reasoning is marked by the use of *if ... then* (I: 08; II: 92). Such reasoning was sometimes directed on conditional reasoning about number (II: 57; I: 90). One case is a good example of reversible reasoning based on knowledge of physical relationships (II: 39).

MODAL   More than one fifth (23%) of the incorrect responses had modal reasons subdivided into three categories:

(a) pseudo-modality

- *because if you put one in there and not one in there, it wouldn't have been fair* (I: 04)
- *no, that won't be fair on that one* (II: 95).

In both cases, the reasoning was pseudo-modal in virtue of an assumed belief that it is not fair for two containers to be treated unequally.

(b) modality of action

- *because if I got three of the orange ones and put them in and I couldn't remember what the number I had, I could pick up that number and put it in, but it wouldn't be the same amount* (I: 01)
- *you could take one out and then they would be equal* (II: 04)
- *you don't have to add one, you could just pretend* (II: 13)
- *because you could put any number you like in* (II: 15)
- *well you can cheat them if you want* (I: 27)
- *no, same as sweets, somebody might not want as much as the other person* (II: 35)
- *because we could take one, and then they won't be the same* (I: 39)
- *because you don't have to put two in each hand* (I: 51)
- *because you can do what number you want to* (II: 51)
- *because I can take them out* (I: 60)
- *because you could have done them at the same time and it would have been equal* (II: 67)

- *I might have accidentally picked up two* (I: 70)
- *you could have got a green cat and put it in there with the red cats* (I: 78)
- *you could make it a secret, like no one seeing, you could get two in each* (I: 88)
- *because all you want to get the same, all you have to do is get all of them out and start again* (II: 89)
- *if you wanted, you could just use one pot and put the other away, and then you could bring it back and that would have more* (II: 91).

These children gave modal reasons of their incorrect response. The common feature concerned the several ways in which they believed that an alternative (possible) action could make, or could have made, a difference. Thus *you could put any number you like in* (II: 15), *you can do what number you want to* (II: 51), *you can cheat them if you want to* (I: 27). Other children reasoned about what has to be done, for example *you don't have to add one, you could just pretend* (II: 13); *it doesn't have to be, because you can do anything you want with them* (I: 47). In some cases, modal reasoning was complemented by analogical reasoning, such as sharing sweets [candies] (II: 35), or by making it a secret and so acting by deception (I: 88).

(c) modality of thought

- *it can be like that, or it can't be* (I: 06)
- *they might be the same, but if you pretend to drop one out* (I: 18)
- *well, it could be the same, but it doesn't have to* (I: 19)
- *because they could be different* (I: 40)
- *it doesn't have to be, because you can do anything you want with them* (I: 47)
- *they can be the same, or they can be less, because you could take two away* (II: 62)
- *the oranges could have had two in the head start* (II: 63)
- *must it be the same? It can be* (I: 67)
- *because it just might be different* (I: 71)
- *there could or there couldn't be* (II: 72)
- *one could have more, but I don't think it will* (I: 73)
- *because if you had them all on your desk and your mother shouted, and then you just stood up and knocked them all off, some could have landed in the bin* (I: 74)
- *because the numbers don't have to be the same* (II: 85)
- *because if one's got less and one's got more, it doesn't have to be the same size* (I: 89).

In each of these cases, the reason was the attribution of a modal property to the object (content) of the child's belief. Since their responses were incorrect, all of these cases amounted to modal errors (Smith, 1993, sect. 25.2). In three cases this was a false-negative denial of necessity, namely a denial that there had to be an equality in study I (I: 47; I: 89), and a denial of an inequality in study II (II: 85). Although one child elaborated this denial by pointing out that *you can do anything you want with them*, this appeal to a modality of action is the basis for a conclusion about a modality of thought, namely that equality is not necessary. Most of these reasons were a false-positive assertion of possibility, namely the assertion that inequality was possible (in study I) or that equality was possible (in study II). The children argued that *it can be like that, or it can't be* (I: 06), *there could, or couldn't be* (II: 72), thus realizing that a (possible) proposition ($\Diamond p$) and its (possible) negation ($\Diamond \neg p$) are co-possible. Some children realized that a necessity is at least a possibility (I: 19; I: 67) in line with standard modal definitions, i.e. ($\Box p \Rightarrow \Diamond p$) or *p*'s necessity entails *p*'s possibility (Sainsbury, 1991). These children were alive to the indefinite ways in which there might have been alternatives to what was the case. Sometimes this was expressed as a bold generality *because they could be different* (I: 40) or *it just might be different* (I: 71). Sometimes this was expressed realistically in that *the oranges could have had two in the head start* (II: 63), or as a human liability to accidents when your mother shouted *then you just stood up and knocked them all off, [and so] some could have landed in the bin* (I: 74).

Recurrence task — modal criterion Q2: reasons for correct responses   Overall, about two fifths (43%) of the responses were correct.

NO REASON   About one tenth (11%) of the responses were given no justification.

EMPIRICAL   Empirical reasons (13%) included:

- *because we started off with one more and that had none* (II: 02)
- *because when you put one in, you forgot to put one in there* (II: 17)
- *[beaming smile] because I want to* (II: 23)
- *because you have been telling me to put those in there* (I: 25)
- *because you are sharing them out* (I: 29)
- *because I put them in at the same time* (I: 36)
- *because you won me, so you should be winning because you've got one more than me* (II: 38)
- *because there was a cat in that one* (II: 41)
- *because I keep putting one in, they need to be the same* (I: 76).

Most of these reasons were reports based on observation or recall, such as *because I put them in at the same time* (I: 36), or *because there was a cat in that one* (II: 41). There were also distortions such as *because you are sharing them out* (I: 29) in as much as the child, not the investigator, did the sharing, or the false suggestion that the investigator had forgotten something at the outset of study II (II: 17). Reasoning by analogy was also evident in as much as the task was interpreted as winning a game (II: 38). One child displayed a paradigm case of egocentric reasoning — that is, factual thinking in the optative (Piaget, 1945/1962) — in the simple declaration that the facts are *because I want to* (II: 23).

NUMERICAL   Numerical reasons (4%) included:

- *because that one's got 12, and that one's got 12* (I: 17)
- *because I counted them* (II: 41)
- *because I have put [counting] 1, 2, 3, 4, 5, 6, 7 and 1, 2, 3, 4, 5, 6, 7* (I: 50)
- *because if you put six in one and six in the other, that means you'll get another six as well* (II: 100).

Some children were reliant on their ability to count (II: 41; I: 50). Other children remarked on the exact number in each container (I: 17). One child believed in a plurality of sixes by suggesting that if there is a six in one container and six in the other *you'll get another six as well* (II; 100), in line with a fallacy attributed by Frege (1884/1950, sect. 38) to some adult logicians.

DEDUCTIVE   Deductive reasons (6%) included:

- *there does because if I took one of the greens away, there wouldn't be more* (II: 20)
- *because if there was the same in there before and you just add another one in each box, there would be the same again* (I: 24)
- *if you put one in there and none in there, that's got more* (II: 48)
- *because if you get two of the same amount all the time, then it makes the same number* (I: 52)
- *because the greens are like Go and the reds are like Stop. But if you add two to those, there will still be more, and if you add one to those, there will be less* (II: 52)
- *because I put them in at the same time, I didn't take any out or I didn't put one in when I didn't put the other in* (I: 56)
- *because if there hadn't been one in there, it would be the same* (II: 76)
- *when you add one on each time, they both get one then two, they both get two. If you do that, does it have to be the same? You don't have to, but it is* (I. 93)
- *if you had four in one and two in the other, and then you swapped them over, and then you swapped them again, and the orange has two and the green has four, then they won't be the same* (II: 93).

A common feature in these cases was a conclusion about what is the case, such as *there does* (II: 20), or *it makes the same number* (I: 52). This was especially clear in one argument which is deductive but non-modal, deductive because the addition of an equality to an assumed equality generates an equality, but non-modal due to the explicit denial of this necessity in the claim *you don't have to but it is* (I: 93). What is the case and what has to be case are clearly separated here, and so it is one thing to make a correct inference and something else again to realize its necessity (Smith, 1993, 1997, 1999d). Other children ingeniously used a *reductio* argument. One child argued in study II that if one was removed, the inequality due to the head start would be eliminated (II: 20), whilst another generalized this argument adding a non-modal twist (II: 93). Other children were reliant on their version of the Euclidean axiom about the equality of addition (I: 24; I: 52). Analogical reasoning was also present in the comparison with traffic lights, augmented by a precise (numerical) quantification (II: 52).

MODAL   Modal reasons (11%) fell into three categories:

(a) pseudo-modality

- *they don't need to have more but they have just got more* (II: 20)
- *because that would be fair* (I: 42).

The reasoning in these cases was pseudo-modal in view of the reference to needs (II: 20) and fairness (I: 42) both of which were inappropriate.

(b) modality of action

- *because when you count them . . . you can add them up* (I: 17)
- *because you have to get one so it'll share* (I: 26)
- *I can't explain, I can't think of anything else* (II: 47)
- *because you can't make a mistake and just go like that [put two in one and one in the other]. You just can't do that, you can just pick them up and pop them in* (I: 87).

One child explicitly denied a capacity to explain the basis of a correct response in a modal claim about action, *I can't explain* (II: 47). Other children identified either alternative actions which you can (I: 17) and can't do (I: 87), or necessary actions such as *you have to get one so it'll share* (I:26).

(c) modality of thought

- *there has to be more in the greens, there doesn't have to be more in the greens, but the greens will always have more in until you have finished* (II: 04)
- *it could be the same, or it doesn't have to be the same, but it probably will have to be the same* (I: 06)
- *there's got to be more in the green, because there was one in there before and none in the orange* (II: 10)
- *because it has to be more because I think we've got more in there* (II: 12)
- *it has to be because we added on more and then I counted them* (II: 15)
- *it can go like that [and] put that in there and that in there — it will be equal . . . because I've got one in each hand* (I: 21)
- *[after denying the equality] there's got to be the same otherwise there wouldn't be an equal amount in each pot* (II: 21)
- *they wouldn't be able to have more in the orange because we started off with none in that pot and one in the green pot* (II: 28)
- *because they have to have the same in each* (I: 29)
- *because we still left one you put in at first in. There has to be more in the green. I kept on doing it [so] it doesn't really make a difference* (II: 32)
- *because that's the way it has to be. From the beginning I started adding more up, so there must be more in the greens* (II: 32)
- *there would have to be more in the orange because we started off with none in that pot and one in the green pot* (II: 32)

- *because it always has to be the same, if you started off with one and you didn't in the orange* (II: 56)
- *it could be something else, cos somebody might get two in that one and one in there* (I: 65)
- *because you gave the greens a head start, so there must be more in there than in there* (II: 87)
- *because in a running race, if a cat and another cat have a race, the green sneaks off and then the cat runs its fastest, then the greens win. So if the greens get a head start and the oranges don't, then there can't be the same in each* (II: 88)
- *because they all have to win at the same time* (I: 92).

In one case, a modal reason about an alternative possibility was inconsistent with a correct response about necessity (I: 65). A comparable inconsistency was made by one child in study I (I: 21). But this same child in study II — also inconsistently — expressed the tautology (*there's got to be the same otherwise there wouldn't be an equal amount*) for the correct response of an inequality. Some children used modal reasoning in expressions of hesitation (I: 06; II: 12), and even in an explicit change of mind about what does and does not have to be the case (II: 04). Other children made necessity an explicit element of their reason (I: 29). Quite simply, *it always has to be the same* (II: 56), either on deductive (II: 10), or numerical (II: 15), or empirical (II: 87; I: 92) grounds. One child reasoned by analogy, combining this with a subtle *reductio* argument in the elimination of the equality of "joint winners" (II: 88). The modal reasoning of another child was spectacularly successful in being both modal and deductive for each of the three modal questions in study II. As he succinctly put it, *that's the way it has to be* (II: 32).

Chapter 7

Discussion

The main conclusions of this study are summarized and then discussed through the three hypotheses in Chapter 6.

(i)   Correct responses predominated over incorrect responses in both versions of the Recurrence task in line with the base criterion of mathematical induction with a significant improvement in correct responses by the children in school Year 2 compared with Year 1.

(ii) Correct responses were more likely than incorrect responses in both versions of the Recurrence task in line with the recursive criterion of mathematical induction with a significant improvement in correct responses by the children in school Year 2 compared with Year 1.

(iii) Incorrect responses were more likely than correct responses in both versions of the Recurrence task with regard to the modal criterion of mathematical induction without evidence of a significant improvement by the children in school Year 2 compared with Year 1, though the overall incidence (46%) of correct responses was greater (48%) in the case of the children in school Year 2.

(iv) Almost all (92%) responses were based on reasons — one fifth with expressions of engagement or delight — which were placed in five categories (no reason, empirical, numerical, deductive, modal) applied to both correct and incorrect responses by children in both school Years.

(v)   There was an asymmetry with regard to the modal criterion in that correct modal responses on the Recurrence task were usually based on non-modal reasoning in contrast to incorrect modal responses which frequently were based on modal reasoning.

(vi) Reasoning on the Conservation task was differentially interpreted in terms of the principle of identity by children making incorrect and correct responses.

(vii) Children who know how to count on the Same task did not always know when to count and when to reason on the Count task, where there was a significant improvement in response rate by the children in school Year 2, nor how to combine counting and reasoning on the Teams task, where there was no evidence of difference due to school Year.

These conclusions can now be related to the three hypotheses. Conclusions (i) and (ii) provide direct support hypothesis (a) which was based on Inhelder and Piaget's (1963) conclusion that children aged 5–7 years can reason by mathematical induction.

Conclusions (iii) and (v) provide indirect support for hypothesis (b) which was based on Inhelder and Piaget's conclusion that these children understand the necessity of this inference. Conclusions (vi) and (vii) provide general support for the independence of (number) reasoning and counting by children during school Years 1 and 2. Each of these is now discussed in this section along with methodological conclusion (iv) in the next section.

7.1 Hypotheses

The third hypothesis concerned the independence of knowing how and when to count. The children in both school Years knew how to count. Performance was at a ceiling level on the Same task, high on question 2 of the Counting task in both studies, and also high on question 1 of the Teams task. Even so, performance was maximal only for the Year 2 children and then only for question 2 in study I on the Counting task. In other words, these children drawn from school Years 1 and 2 did not count successfully and reliably, even though the numbers in the arrays were small (4 or 5; 11 or 12). Knowing how to count is not the same as knowing when to count and these children did not always know when counting was appropriate. This was evident in two different ways: they sometimes did not count when this was appropriate; and they sometimes counted when this was not appropriate.

On the Counting task, the children did not always realize that counting was appropriate for question 1. The difference in performance between the two questions was significant for each Year other than the Year 2 children in study I alone. In other words, the children tended not to count in answering question 1 when counting was appropriate. Further, there was no significant difference in performance due to Year on this same task in study II, where the success rates on both questions were lower. The same tendency was evident on the Teams task. Performance approached ceiling levels for question 1 and there were no significant differences by Year, neither overall nor for each of the three specific versions of this question. Yet there was a significant difference in performance between the two questions. Although these children were able to count, they tended to count the objects open to observation in front of them rather than to count as directed the objects specified under the hypothesis in the question put to them. Miscounting due to misconception was also apparent notably on question 2 of the Conservation task where some children defined the lines using arbitrary spatial criteria.

The contrary tendency was also apparent in that the children counted when this was not appropriate. This was apparent in relation to both the base and recursive criteria of the Recurrence task where a quarter of the cases were regularly due to numerical reasons based on counting. Interestingly, this was not so in relation to the modal criterion on this task. Indeed, the reasons for incorrect responses amounted to about 4 percent of the cases. The children did not seem to realize that counting is only one way to answer a numerical question. Although some of the children did reason in line with Euclidean axioms, many other children did not do this. The child in school Year 2 who counted one by one the 18 cats in each of two containers after adding a cat to each container, agreed by him to be equal on the basis of his prior counting of the 17 cats in each, displayed accuracy and persistence in this display of numerical expertise. But at what expense of an ability to reason! Other equally persistent children with a flair for inaccuracy reckoned that the contents were no longer equal in a context where one more had just been added to the equal containers by themselves. The same tendency was also apparent in responses to question 2 of the Conservation task in that some children regarded counting as the sole way to ascertain that the number was the same. As one Year 1 child put this:

● *I don't know because I haven't counted them.*

In general, this evidence is compatible with the third hypothesis in that these children had learned that counting is one way to give correct answers to questions about number. But they did not always realize that questions about number are diverse. Some questions concern "How many?" and counting may be an appropriate way to answer such questions. Other questions concern "As many?" The crucial insight which was exploited in the logicism of Frege and Russell was that reasoning about "as many" is both independent of, and also superior to, counting on the basis "how many". It is for this reason that an expertise which combines good answers to either question as appropriate is both desirable and necessary for intellectual advance. Although the ability to count is acquired during the preschool years, it is not always used appropriately by children during the early years of schooling.

There are three ways in which this hypothesis can be interpreted. One is that the acquisition of knowledge is not the same as its use and children experience difficulty in learning how to fit their knowledge to relevant contexts (Bryant & Kopytynska, 1976; Sophian, 1995), especially in view of the optimizing or constraining effects of different social settings (Nunes & Bryant, 1996, 1997). Inhelder (1954/2001; Vonèche, 2001) evidently had this view. A second position is that the early possession of one ability interferes with the development or use of another ability (Gelman & Williams, 1998). A third position is that mathematics learning in schools is culturally linked with learning how to count. This sets the standard for all mathematics learning, including learning on reasoning tasks. In consequence, the ability to count is used in default, unless it is specifically overridden. Each of these interpretations has the same implication, namely that learning how and when to reason is an alternative way of gaining certain types of knowledge. Reasoning skills have a further benefit in virtue of their indispensability to mathematical progress.

The first hypothesis concerned reasoning by mathematical induction. By implication, the findings from the present study are incompatible with the Frege–Russell position in which mathematical induction is reduced to logical deduction, but compatible with Poincaré's position in so far as mathematical intuition is interpreted in terms of iterated action. But it is worth restating one of the main points made by Frege along with a constructivist interpretation.

Central to mathematical induction in this study was reasoning about numerical equality or inequality. It was pointed out in Chapter 2 that equality is identity. The credit for pointing out that this notion is as subtle as it is vexatious in the minds of philosophers is due to Frege (1884/1950; 1892/1960). Frege's point was well taken by Piaget, probably whilst he was still at school in Neuchatel (Smith, 1999a). At issue in Piaget's theory was the question "How do children understand equality, for example in their reasoning about the conservation of number?" To say that *number is conserved* means that something in the reasoning is the same, i.e. is self-identical. To make the same point: if something at the outset of the reasoning is the same as something at the end, and if that something is a number, then *number is conserved* means that the number at the outset is equal to the number at the end. Piaget's conservation tasks were designed to ascertain what — if anything — is the same in children's reasoning. Note that if children believe that there are "two things" which stay the same in the task, by definition these two things cannot be self-identical. Two lines cannot be one and the same line. In the actual world open to observation, everything is what it is and not another thing. But numbers are *not* things in the actual world, even though current doubts remain as what numbers are (Benacerraf, 1973; Hersh, 1997). Numbers are abstract objects. Note, however, that this is to name the problem of giving a good analysis of abstract objects. It is not a commitment to one preferred analysis such as realism (Katz, 1995). This is because nominalism (Field, 1980) and constructivism (Dummett, 1991) have to be reckoned with as well.

Although Piaget (1950, 1967b) was committed to constructivism, there is no convincing argument on offer as to a constructivist ontology. What is on offer here is a constructivist epistemology (Smith, 1999e). According to Inhelder and Piaget (1979/1980, p. 21), their constructivism was directed on the "creation of new objects (*êtres*) such as classes, numbers, morphisms etc." during ontogenesis and in the minds of children. Their research programme included the well-known studies of conservation (Piaget & Szeminska, 1941/1952; Piaget *et al.*, 1948/1960). And it included as well the much less well-known study of reasoning by mathematical induction (Inhelder & Piaget, 1963). Note that reasoning by mathematical induction generates the problem that is central to number conservation, namely whether and how children understand that something is the same (self-identical, equal to itself) in a chain of reasoning. Although spatial transformation was implicated in the design of Piaget and Szeminska's tasks, temporal conservation was the dominant variable in the Recurrence task in virtue of the repeated additions which the children carried out (see also Bryant, 1995, 1997).

There was a significant difference in performance between Years 1 and 2 on the Recurrence task. Even so, the overall rate of successful responses in each Year was high for all of the questions relevant to both the base and recursive criteria of mathematical induction. Since this evidence shows that both criteria had been met, it is compatible with the first hypothesis, that children do reason by mathematical induction on the basis of iterative action. This interpretation of the evidence is psychologically distinctive. During ontogenesis, the development of knowledge is based on children's own activities. In the Recurrence task, the activity was simple, namely simultaneous addition with the same quantity in each hand. The action was physical, amounting to transferring — one at a time and one in each hand at the same time — the physical objects from two piles to two containers. Their reasoning shown by the inferences which the children drew from their own activity was another matter. Across the sample, these inferences were varied and complex, varied because different inferences were drawn and complex because not all of the children realized that abstract objects — and not merely actual objects — were at issue. The children did of course know:

● *there's loads of different numbers you could get.*

But they did not always realize — or if they did, they did not always display this realization — that each natural number is a unique abstract object. There is only one number 1, only one number 2, and so too for each and every number. Some children suggested that their transfer had resulted in an equality since *the same numbers were in each container.* This must be false since an equality means that there is one and only one number in their reckoning. Other children denied the necessity of this equality because *the numbers don't have to be the same.* The use of the plural *numbers* is significant. These children are regarding number as a physical object or physical property, such as height or colour. The fact that they had added 15 cats to each container was interpreted by these children as the acquisition by the container of a new property. So interpreted, different containers do not have to have the same number in each since their properties are regarded as different in the first place. This is a fundamental mistake (Frege, 1884/1950). It is also an understandable mistake in view of the complexity of the notion that number is a property of abstract objects (Benacerraf, 1973; Katz, 1995).

Following Piaget (1936/1953; 1970/1983), several developmentalists (Langer, 1980, 1998; Müller *et al.*, 1998; Smith, 1998a) have identified action as the basis for the development of knowledge. There are, of course, many types of action (physical, social, cultural, moral) and so the physical activities which were central to Inhelder and Piaget's (1963) study should not be regarded as the sole type (Piaget, 1977/1995). It is an open question whether individual or social actions contribute to intellectual development, interpreted as the construction of abstract objects (Inhelder & Piaget, 1979/1980).

Even though the children's reasoning by mathematical induction was based on their actions, there was clear evidence that some children used deductive reason expressed as an "*if . . . then*" conditional. In some cases, this reasoning was sound and systematically carried through:

● *if you put 6 cats and 7 cats in, it won't be the same, but if you put 7 cats in and 7 cats in, it would be the same.*

In others, it was used creatively to express some version of the Euclidean axiom about the equality of addition:

● *if there was the same in each and you added one more in each, it would be the same.*

This evidence is persuasive since these children were reasoning autonomously in a deductively appropriate way. It is compatible with the position that children know how to reason conditionally in counterfactual contexts (Hawkins *et al.*, 1984; Johnson & Harris, 1994). Yet there is contrary evidence and interpretation as well with the reliable finding that conditional reasoning based on inference rules is typically accomplished in early adolescence (Markovits, 1993; Markovits *et al.*, 1996). The children in this study did not perform well in the counteractual context of the Teams task. Nor was there a check that these children could relate a conditional with its negation (Smith, 1993, sect. 24). This means that the use of the expressions *if . . . then* and an understanding of the logic of conditionality may not amount to the same thing. During ontogenesis, a conditional expression can be understood as a conjunction (Smith, 1987; Müller *et al.*, 1999).

One way to combine both positions is under the proposal that conditionality is an element in an abstract object. The structure of this object is well known and specified in logic texts (Lemmon, 1966; Sainsbury, 1991; Wittgenstein, 1972, para. 6.36111). Piaget (1949) invoked this structure on the grounds that the serial acquisition of individual elements in it can lead to the eventual mastery of the complete system. His proposal was that the distinction between conditionals and conjunctions can be understood in many different ways, but the logical distinction is complex. This proposal can be understood through a key distinction due to Reichenbach (1947) who pointed out that propositional operators (such as conjunction, conditionality, disjunction) can be understood bidirectionally. By this he meant both that a propositional operator can be understood through its truth conditions set out in a truth table, and that these truth conditions set out in a truth table can be understood through a propostional operator. In logic texts, these are logically equivalent. According to Piaget, there is a unidirectional understanding during ontogenesis in that the second alternative in Reichenbach's distinction is psychologically primitive. Thus in Figure 7.1, the conditional relationship in row **0** is constituted by the conditional operator ($\rightarrow$) which is different from the operator for conjunction (&). The difference is shown in the truth conditions set out in the relevant truth table under column **2**. The entries to columns **1** and **3** are the same for both conditionals and conjunctions, where the entry 1 means that there is a case corresponding to *p* (or *q*), and 0 means that there is not. The truth tables differ under column **2** where the four entries are different for conditionals and conjunctions. Under Piaget's proposal, children acquire an understanding of the parts before the whole (abstract structure). These parts correspond to rows. Each row can be understood individually without regard for the other rows. All four rows can be understood jointly through columns **1** and **3** but without regard for the overall structure in column **2**. A unidirectional understanding becomes bidirectional if and only if column **2** is used in the interpretation of the four rows and other two columns (see Smith, 1993, sect. 24.2). Just such a position was restated by Piaget (1977/1986; cf. Piaget & Garcia, 1987/1991) as the claim that there are *ilôts* (islands) of understanding during intellectual development due to the optimizing possibilities of powerful learning contexts.

The second hypothesis concerned modal reasoning about "what has to be". There was no difference in the children's performance by school Year with regard to the modal questions on the Recurrence task, although the overall incidence of modally correct responses was higher than would have been expected from the research literature with its implication that modal reasoning is absent from childhood. A similar

	p = there is red wine on the table					
	q = there is white wine on the table					
Row	**Column**			**Column**		
	1	**2**	**3**	**1**	**2**	**3**
0	p	$\rightarrow$	q	p	&	q
1	1	1	1	1	1	1
2	1	0	0	1	0	0
3	0	1	1	0	0	1
4	0	1	0	0	0	0

Reichenbach (1947): bidirectional understanding of columns and rows in logic

- columns can be understood through rows
- rows can be understood through columns

Piaget (1949): directional understanding during ontogenesis

- columns are initially understood through rows

Figure 7.1. Reichenbach and Piaget on the understanding of prepositional operators and their truth conditions.

finding was apparent on the modal question of the Conservation task. So these responses were not compatible with the main hypothesis in which a difference due to Year was predicted and to that extent gave no support to the second conclusion drawn by Inhelder and Piaget (1963). But the children's reasons led to a different set of findings. Since there was no statistical analysis of these reasons, no generalizations should be drawn. Nonetheless, these findings set out examples — "good specimens" — which are available for interpretation.

There was a clear asymmetry in that modally correct responses were rarely based on modal reasons, even though good modal reasons were often given for modally incorrect responses. Modal reasons for incorrect beliefs were based on real-world understanding

- *because if you had them all on your desk and your mother shouted, and then you just stood up and knocked them all off, some could have landed in the bin*

or on mathematical reasoning

- *because if one's got less and one's got more, it doesn't have to be the same size.*

Non-modal reasons for correct beliefs were predominant, and these reasons were applicable to the truth-functional question ("Is there the same in each or not?") rather than to the modal question ("Does there have to be the same in each or not?"). Examples included reference to the action of adding

- *because I put them in at the same time,*

numerical reasons

- *because I counted them,*

as well as deductive reasoning

- *because if you get two of the same amount all the time, then it makes the same number.*

Cases of modal reasons for correct beliefs were rare but they included

- *because we still left one you put in at first in. There has to be more in the green. I kept on doing it [so] it doesn't really make a difference.*

Further, pseudo-modal reasoning was sometimes displayed by these children, both for modally incorrect responses

- *no, that won't be fair on that one,*

and also for modally correct responses

- *yes, because that would be fair.*

A surprising finding was the autonomous display of modal reasons given by some children in their replies to non-modal questions. There were several cases where this occurred, including incorrect beliefs about the addition of "any number"

- *that might have sixty-six and that might have sixty-three — they're not the same number,*

and also for correct beliefs about the addition of "any number"

- *if it's the same, it's got to be the same.*

In short, modal reasoning, which is usually confined to adolescents (Morris & Sloutsky, 1998; Moshman, 1998) or even to adults (Bell & Johnson-Laird, 1998), was evidently present in some form in these children's reasoning. But this evidence is compatible with other accounts of the development of modal reasoning, one due to Piaget and the other to Piéraut-Le Bonniec.

The 1963 study of mathematical induction was invoked by Piaget (1977/1986, p. 303) in his essay on necessity. This reference was combined with the claim that

> at all levels inference has the role of accounting for local necessities.
> This is so [when] local necessity is at issue, arising well before the

constitution of operational structures in the relations between schemes, however elementary. (p. 307)

This is a clear statement that modal reasoning based on the inferential links between action-schemes is developing in advance of concrete operational thought. From this perspective, the child is an agent whose own activities are generative of meaningful experience in the construction of knowledge. This experience falls into two types based on an agent's self-regulation of action as manifest in abstractions and generalizations drawn from those actions. Piaget (1977, pp. 39–40) named these two types physical and logico-mathematical experience. An alternative characterization was empirical and reflective abstraction (Piaget, 1977/2001). Yet another was inductive and logico-mathematical generalization (Piaget, 1978, p. 220). This plethora of distinctions could be viewed as a semantic fudge to hide the absence of a developmental mechanism (Boden, 1979; Flavell *et al.*, 1993; Klahr, 1999; Siegler, 1996) just because naming is not the same as explaining when analysis and corroboration are required as well. Yet there is a fundamental distinction here and this distinction has long been acknowledged in philosophy from Plato (1941) through Kant (1933) to Wittgenstein (1972). The actual world is only part of reality. It is therefore a fair question to ask whether an epistemic instrument (observation, empirical inductive learning, counting) which may produce good knowledge of the actual world is also productive of good knowledge of other parts of reality. Knowledge of number is a case in point. Piaget's several distinctions are due to his recognition of the subtlety of this problem. In the present study, some version of this distinction may be implicated in these two different responses

- *the numbers are the same*
- *it will have to be the same number.*

The former refers to the plural *numbers*. The latter is a singular reference to *number*. This difference is fundamental. As a reference to the infinity of natural numbers, the former would be fair enough. Yet this child's reference was to a number (the number was six) believed to be true of the contents of the containers. Any individual number is unique and is in any case not part of the actual world. The latter reference is the preservation of an equality through serial addition to the containers. The singular reference is essential to this realization, namely that whatever the number is, it is and has to be the same. It is this distinction which is at the forefront of Piaget's account. Granted that all learning has its starting point in the actual world of objects with causal properties, how in fact do children develop an understanding of these abstract objects and their normative properties? It is this question which is taken up in the next section.

7.2 Developmental Mechanism

The proposal is that a mechanism of developmental advance is primarily due to an interplay between an agent's activities in response-making and reason-giving. This is a special case of the advance from action to thought. This case fits the reconciling unit of analysis which combines "causal facts" and "normative facts" in acts of judgment. Five epistemic norms are set out in an AEIOU framework.

Questions about the advance from the causal to the normative are questions which are as fundamental as they are unexplained (McGinn, 1991; Nagel, 1997). Such questions were central to Piaget's research programme. This research programme was set out in Piaget's (1918, p. 116; see Smith, 2002a,b) first book. Its standard question was the advance from fact to norm, from the causal to the logical (Piaget, 1977/1995, p. 51). A paradigm example was the construction of necessary knowledge. There were repeated statements of this problem:

> the necessity resulting from mental experiment is a necessity of fact; that which results from logical experiment is due to the implications existing between the various operations. (Piaget, 1924/1928, p. 237)

> the main problem of any epistemology is in fact to understand how the mind succeeds in constructing necessary relationships, which appear to be independent of time, if the instruments of thought are merely psychological operations that are subject to evolution and are constituted in time. (Piaget, 1950, p. 23)

> the emergence of logical necessity constitutes the central problem in the psychogenesis of logical structures. (Piaget, 1967b, p. 391)

> necessity is not an observable, based on a reading from objects. Necessity is always the result of constructions inherent in a subject . . . From this arises our interest in the study of its formation during psychogenesis. (Piaget, 1977/1986, p. 302)

> [the problem concerns] this transition from a temporal construction to an atemporal necessity. (Piaget & Garcia, 1983/1989, p. 15)

Necessary knowledge is atemporally true. This is a normative property of knowledge. But if such knowledge is constructed, this occurs during time and so the construction is temporal. How, then, does this take place? Some version of this problem was implicated in the epistemologies of Plato, Kant and other philosophers. What Piaget set out to do was to investigate this problem in an empirical and developmental epistemology. In this epistemology, the question became "How does necessary knowledge develop?"

This question generalizes to other epistemological norms (see Box 6.2). Examples of such norms are given in an AEIOU framework: autonomy, equality/entailment (modal knowledge of necessity), intersubjectivity, objectivity, universality. In psychological studies of intellectual development, a central question is the causal contribution made by context to human learning in arithmetic (Nunes & Bryant, 1996, 1997). In a developmental epistemology applied to intellectual development, this focus would be augmented to include normative contributions such as the five equally interesting, epistemological properties (Smith, 1999d, 2002a,b). Along with the example in Box 6.2, these can now be explained using the Kantian example: $7 + 5 = 12$.

- autonomy

Numbers have normative, not causal, properties. It is one thing for there to be compliant acceptance that $7 + 5 = 12$ due to the potency of causal context. It is something else again for an equality to be accepted through reasoned assent (Piaget, 1967/1971, p. 49).

- equality (entailment, necessity)

The equality $7 + 5 = 12$ is a necessary, not empirical, truth. There are no exceptions, because there could not be any. It is one thing for this truth to be understood to be correct. It is something else again for its necessity to be acknowledged as something which could not be otherwise (Piaget, 1977/1986, p. 312).

- intersubjectivity

The equality $7 + 5 = 12$ is one and the same equality open to all, and so has the potential to be common ground between different thinkers (Piaget, 1977/1995, p. 94). It is one thing for this truth to be represented in an individual way which is different for different individuals. It is something else for this to be the self-same thought of different thinkers.

- objectivity

This equality is a true equality, and so is not false. This difference is inherent in knowledge in that only truths can be known (Piaget, 1967/1971, p. 35). It is one thing for someone to make a response which is true, for example to say "12" to the question "What is $7 + 5$?" It is something else to acknowledge this to be true and to regard it as true.

- universality

This equality is always the case, and as such is one basis for mathematical generalization. It is one thing for the equality $7 + 5 = 12$ to be accepted in a range of contexts. It is something else for its truth to be understood to be always true even in one context (Piaget, 1968).

These five epistemological norms are merely five examples which do not exhaust the class. They are applied to education in Chapter 8. There are other norms in the reckoning, including values (Donaldson, 1992), self and personality (Demetriou & Kazi, 2001), and human motivation (Brown, 2001). They can be interpreted through a reconciling unit of analysis in terms of an act of judgment (Smith, 1999b,c, 2002a,b). Any act has causal properties for psychological investigation. Any judgment has normative properties for epistemological investigation. The problem central to Piaget's research programme is the advance from one to the other. And this advance requires a mechanism.

The proposed mechanism is in terms of action in that action is a combination of causal and normative properties:

> action necessarily deforms the ideal in virtue of its mixture of fact and
> norm. (Piaget, 1918, p. 116)

Behind this remark are two questions which can always be asked about any action: "Was it in fact performed?" and "Was it the right action to perform?" What Piaget realized was that if actions have both causal and normative properties, the agent is in a position to develop normative knowledge on the basis of actions on causal objects in the actual world. This would secure a link between action with its biological basis and knowledge with its epistemological properties. In this way would it be possible for agents to advance from causality to normative implication, from action to conscious necessity (Piaget, 1977/1995, p. 53). Any action is a causal fact for explanation in psychology. Any action is also a normative fact for explanation in (developmental) epistemology. Under this proposal, intellectual development is an advance in two respects. One is the advance in causal understanding of objects in the actual world. The other is the advance in normative understanding of abstract objects which are equally part of reality.

   Piaget (1936/1953, p. 240; 1977/1995, p. 56) has long argued for a *décalage* between action and thought. In a constructivist model, epistemological norms would be neither innately available nor acquired through socio-cultural interaction, but would instead develop over time. According to Piaget, there is a *tertium quid* (third alternative) which he reckoned is equilibration (for a critical discussion, see Smith, 2002a,b). So the main question concerns not whether or when but how this takes place. In such a model, action is the source of knowledge:

> knowledge is constantly linked with actions or operations [and so] at its
> origin neither arises from objects nor from the subject, but from inter-
> actions — at first inextricable — between the subject and those objects.
> (Piaget, 1970/1983, p. 904/p. 104)

This account provides a possible mechanism for the understanding of both equality and necessity. The empirical experience arising from the Recurrence task is the action of repeated addition. This amounts to different action-tokens of the same action-type. Thus identity is secured. Further, an empirical induction can be drawn about the regularity of the outcome of this action: it was the same before, so it is the same now. The logical experience due to the Recurrence task is one-and-the-same action directed on one-and-the-same object. It will be asked "One what?" There are two answers here. First, the answer could be "One plastic cat". Second, the answer could be "One". The distinction here could well match the different positions taken by the children above:

- *they are the same numbers*
- *it will have to be the same number.*

The use of the plural *numbers* in the first of these can be interpreted through observational learning and empirical induction. The use of the singular *number* in the second can be interpreted through logical reasoning and mathematical induction.

   Is there a model of the logic of action? There is a general claim to make here, followed by a specific one.

The general claim concerns practical reasoning. Following Aristotle (1987) in his *Nicomachean Ethics*, reasoning falls into two types. Theoretical (syllogistic) reasoning is one type which is based on reasons for beliefs, i.e. the propositions expressed in beliefs. Practical reasoning is the other type which is based on reasons for actions. This distinction is fundamental (von Wright, 1971, 1983a). It has the consequence that actions — and not merely beliefs — may be based on inferences and thereby have a logic. Deontic logic is a logic of action which fits practical reasoning (Ross, 1968; von Wright, 1983a). Deontic reasoning concerns directives, rather than propositions, and these directives have a modal character in laying down commands (what should be done), permissions (what may be done) and prohibitions (what is forbidden and so ought not to be done). Although deontic reasoning has figured in recent research, there has been too much concern with the development of (theoretical) reasoning about deontic propositions and too little about deontic directives in practical reasoning (Cheng & Holyoak, 1989; Falmagne & Gonsalves, 1995; Fiddick *et al.*, 2000). The same is true of deontic reasoning in educational contexts. There is scope for further work here, even though this general claim is not otherwise pursued here.

The specific claim starts with Piéraut-Le Bonniec's (1980) work and her argument that modal reasoning has its origin in knowing how to do things. This is a commitment to a model based on modal logic. A central element in her argument is that Piaget's account of formal operations does include modal reasoning but that it is incomplete in two respects. One concerns developmental sequences: are formal operations interpreted as reasoning in adolescence in truth-functional or modal terms during childhood? The other concerns developmental mechanisms: how does modal reasoning makes its way into the reasoning of adolescents? Doubtless, these are both good questions. What is evident from the present study is that Piéraut-Le Bonniec has made a good point in drawing attention to the display of modal reasoning in children's real-world actions. The modal reasoning on display in this study also included the modality of action which was distinguished from the modality of belief. The children were endlessly inventive about the alternative actions open to them. This was notably the case with the reasons for modally incorrect responses:

- *you could take one out and then they would be equal* (II: 04)
- *you could have got a green cat and put it in there with the red cats* (I: 78)
- *well you can cheat them if you want* (I: 27).

Yet these reasons are modal reasons about what is to be (or may be) done. As such, they differ from

- *because it just might be different* (I: 71)
- *well, it could be the same, but it doesn't have to* (I: 19)
- *because if one's got less and one's got more, it doesn't have to be the same size* (I: 89)

which are modal reasons about what must (may) be. This is a fundamental distinction between *tun-sollen* (what ought to be done) and *sein-sollen* (what ought to be) which is well discussed by von Wright (1983a). It is this distinction which was respected in the categorization of the children's modal reasons in Chapter 6.

A potential means of developmental advance can be set out as an interplay between responses and reasons since these are two actions. Making a response is an action which shows what an agent "knows how to do". What is still to be determined is what sort of knowledge this is in terms of its relations with other knowledge in that agent's system. There is a continuum here, ranging from no connections at all (totally disconnected knowledge), causal connections effected by contextual factors, and normative connections in incipient systems of knowledge. If there is knowledge "in action" (i.e. response-making action), this knowledge will have relations with other knowledge, and these relations are clarified through the reasons for the response. This is because reasons are expressible, and in this study were actually expressed by the agent. This means that reason-giving is also an action since this is also something which an agent does. In turn, the same problem recurs. The nature of this knowledge "in action" (i.e. reason-giving action) is still to be determined through its connections with other knowledge in the systems, and notably with the knowledge in the response-making action. In the context set by response-making, what the action of reason-giving can provide is a characterization of their interrelations, and so the extent of their mutual organization. This proposal fits Piaget's model in several ways. First, the proposal fits a psychology of normative facts (Piaget, 1961/1966, p. 153) with a dual focus on "the coordination of factual data and normative validities". This proposal includes both causal and normative elements. Second, it fits Piaget's (1977/1995, p. 51) view about the advance from "the causal to the logical" in the development of knowledge, since a causal organization conferred by reasons on responses and a normative organization are not the same thing. It is through reasoning in the sense of an agent's giving reasons for that same agent's own responses that the advance could take place from one to the other. Third, it fits Piaget's (1975/1985, p. 29; translation amended) definition of a regulation: "in virtue of its very exercise, every regulation moves toward both retroactive and anticipatory effects". This means that reason-giving action is organizational in either of two ways, one retroactive and the other anticipatory. Retroactive organization is typically manifest as causal organization, for example when the children gave empirical reasons for their responses. Anticipatory organization is typically manifest as normative organization in terms that match between what is believed (wanted, hoped, expected) to be the case and what is then found to be the case. Fourth, it fits Piaget's (1978, p. 221) principle that

> constructive generalization does not consist in the assimilation of new content to forms already constituted, but rather in generating new forms and new content, and therefore new structural organizations.

It is in this sense that equilibration is reckoned to be productive of novelty, i.e. the development of new knowledge (Piaget, 1975/1985, p. 3).

Using this proposal, an example can now be clarified in two moves. The first move breaks down into two smaller steps. One concerns propositional logic which is not the same as modal logic. True propositions are alike in being true. Suppose a coat has two pockets such that

> there is £7 in one pocket and £5 in the other making £12 altogether

Manifestly, it is also true that

$7 + 5 = 12$

So both propositions are true. Yet these truths differ in one crucial respect. The latter is a necessary truth: it could not be otherwise. By contrast, the former is not a necessary truth: it could be otherwise. Someone may have stolen the money from the coat pockets; the coins could have been counterfeit; the coins could have fallen through the holes in the pockets. So there is a distinction to draw. And this distinction can be drawn in logic. In logic, a proposition

p
(read: *p*)

is different from

$\Box p$
(read: *p* is necessary)

The latter can be defined as

$\Box p = \neg \Diamond \neg p$
(read: *p* is necessary just in case it is not possible not-*p*)

in that a necessary proposition is one whose negation is impossible. Notice that one modality (necessity) is defined in terms of another (possibility) in the same modal (alethic) family (Haack, 1978; Marcus, 1993; Sainsbury, 1991). In other words, a (true) proposition *p* is such that it could be false. This is because in (non modal) propositional logic, truth and falsity are the sole respects in which propositions differ. In (propositional) modal logic, this difference between truth and falsity is augmented through another as to the mode or manner in which something is true or false. Thus $\Box p$ and *p* are different in modal logic in that the invalid inference

$p \Rightarrow \Box p$
(read: *p* entails *p* is necessary)

is a fallacy: the necessity of *p* is never entailed by the truth of *p* (Haack, 1978; Marcus, 1993; Sainsbury, 1991).

The second of the smaller steps follows on. The way to check a necessity is by the identification of the possibility of its negation. If this is a possibility — if it is possible for a negation of a proposition to be true — then that proposition is not necessary. The denial of the necessity of *p* can be expressed as

$\neg \Box p$
(read: it is not the case that *p* is necessary)

which in turn means

$$\neg \Box \, p = \neg \neg \Diamond \neg p$$

(read: it is not the case that p is necessary just in case it is not the case that it is not possible not-p)

which can be simplified through dropping the double negation as

$$\neg \Box \, p = \Diamond \neg p$$

(read: it is not the case that p is necessary just in case it is possible not-p)

In short, what follows from these modal definitions is a clear-cut way of deciding whether or not some proposition is necessary or not. Is its negation possible or not? If it is not possible, the proposition is necessary. If it is possible, the proposition is not necessary.

This leads to the second move. Many children who made modally incorrect responses were nonetheless sensitive to modality. This was shown by their modal reasons. These modal reasons were reasons for the response

- *it is not necessary that . . .*

i.e. the denial of a necessity. In conformity with their response, these reasons identified the possibility of a negation. Note this consistency. Note that any such reason for the possibility of a negation was in line with their response which is the denial of a corresponding necessity. The children identified many different ways in which this possibility could be instantiated, such as the six ways mentioned a couple of pages back. These reasons are interesting in their own right, testimony to the inventive functioning of human minds in action. They have an added potency in view of this modal consistency in that the modal reason fits the modal response. As such, this leads to the possible mechanism of advance. It is in their reasoning that children may recombine the parts of a complex structure, reorganize one part of an abstract object by reference to its other parts in a complex whole. It was proposed above that this could be responsible for the mastery of conditional reasoning. Exactly the same proposal could be responsible for the development of modal reasoning. This proposal is interpretable as a developmental mechanism in terms of equilibration (Smith, 2002a,b).

The proposal is that modal questioning on the Recurrence task triggered reasoning about alternatives, leading to the belief that "things could be otherwise". The constraint which these children had not overcome is that their actions were directed on actual objects, and they were of course quite right to argue that

- *you can cheat them if you want to.*

But that misses the point. The modal question concerned a hypothesis which the children were invited to entertain, and the hypothesis concerned the addition of an equal number to an already accepted equality (or inequality in study II). The children were physically adding to the containers. But this was not all that they were doing. They were also entertaining a hypothesis about number and the preservation of an equality through serial addition. And number is an abstract object.

On this interpretation of the difficulty, the demarcation of different levels of thought, one level directed on actual objects and another on abstract objects, is a principal problem. This demarcation problem was illustrated in a modal reply which one child made to a non-modal question. Chris (aged seven years) had just made eight additions in study II and so there were nine green cats and eight orange cats in the containers. The dialogue starts with a correct answer to the non-modal question Q1 (see Box 7.1). In justifying this correct response, Chris gave a modal reason — there *must* be more Greens. Intending to exploit this modal realization, the interviewer reformulated this — there *has to* be more Greens. If Chris's reasoning was directed on abstract objects, assent should have been given. Instead, Chris denied this. The denial was based on reasoning about actual objects, namely that another orange cat could be placed in the other container. That is, an intellectual switch took place, i.e. there was an attentional shift from one level of reasoning to another. The presence of attentional shifts was noted in his first psychological papers by Piaget (quoted in Smith, 1993, p. 152) to be recast in terms of levels of reflecting abstraction in his constructivist work (Piaget, 1977/2001).

Box 7.1: Chris's reasoning about "what is the case" and "what must be the case"

Interviewer	Is there the same in each, or more there (pointing to one container), or more there (pointing to the other)?
Chris*	More in the Greens.
Interviewer	Why is that?
Chris	We started with one Green and no Orange and I kept on adding one. So there must be more Greens.
Interviewer	So there has to be more Greens!
Chris	No.
Interviewer	Why?
Chris	'Cause you can put one more in the Orange to make it the same.

* not the child's actual name

In a small number of cases, modal reasons were given for a modally correct response

● *there's got to be more in the green, because there was one in there before and none in the orange* (II: 10)

- *because we still left one you put in at first in. There has to be more in the green. I kept on doing it [so] it doesn't really make a difference* (II: 32)

often coupled with an affirmation which reveals the immense difficulty of extricating "normative reasoning" from a powerful "causal context"

- *there has to be more in the greens, there doesn't have to be more in the greens, but the greens will always have more in until you have finished* (II: 04)
- *it has to be because we added on more and then I counted them* (II: 15).

These modal reasons for modally correct responses are here interpreted as actualizations of the logical experience inherent in these children's actions. As such, they are interpreted here as the outcome of the capacity to demarcate and maintain a difference in levels of reasoning directed on abstract objects in a context set by actual objects.

In short, the proposal is that a mechanism of developmental advance is primarily due to an interplay between an agent's activities in response-making and reason-giving. It is the latter which contains the potential for an advance to be made in the sequel, i.e. that agent's future actions. The same problem then arises about the extent to which the knowledge in any (future) action fits into a system with the knowledge already "in action".

Two objections may arise, one about other causal variables and another about other models of action logic.

One objection is that the study in Chapter 6 was insufficiently sensitive to other causal variables. Piaget and Szeminska (1941/1952, p. 149) famously denied that they had any interest in such variables. But this is not binding on everyone. Thus it is possible that children may have performed differently under different causal conditions, such as those implicated in alternative accounts of number learning which were reviewed in Chapter 4. There are two replies to make here. One is, of course, to agree. This admission is important, and it means that there is something to do in the sequel. But there is a qualification as well which is the second reply. This is a reminder that "causal facts" and "normative facts" are not the same thing. An investigation of "causal facts" concerns the contribution made by (other) causal variables under different experimental conditions. These variables could be expected to influence the incidence of correct responses made by children, maybe too their reasons for the responses. An investigation of normative facts concerns the contribution made by reasons to the responses, whether correct or not, under the operative causal conditions. These reasons are phenomena in their own right and they could be expected to have some normative influence on the responses. Right now, it is sufficient to point out that the design in Chapter 6 was directed on "normative facts" in contrast to the vast majority of studies which are directed on "causal facts".

A different objection concerns other models of action logic. This objection is interesting just because modal reasoning was central to the design in Chapter 6. Such reasoning has been interpreted in this chapter through the alethic logic of modality. But there are other modal logics, including (modal) deontic logic (Ross, 1968;

von Wright, 1983a). The alethic modalities (necessity–possibility–impossibility) are similar to deontic modalities (obligation–permission–proscription). This formal similarity in their definition is well known (cf. Piéraut-Le Bonniec, 1980; Smith, 1997). Also well known are fundamental differences (Ross, 1968; von Wright, 1983a). The *ab esse ad posse* principle (the actual is possible) is valid in alethic reasoning, but invalid in an application to deontic reasoning where it would amount to the principle that the actual is permitted. The inference

some people rob banks

therefore

it is possible that some people rob banks

is a valid inference in alethic reasoning: if the premise is true, the conclusion must be true as well. But the inference

some people rob banks

therefore

it is permitted for some people to rob banks

is invalid: even if this premise is true, the conclusion is an invalid inference in deontic reasoning. But if alethic and deontic modalities are different, maybe the latter could be used, whether interdependently with or even independently of alethic modality. Three replies can be made to this objection. First, the late development of deontic logic should be noted (von Wright, 1983a). It is for this reason understandable that Piaget (1949) sided with the dominant position in logic. Interestingly, this dominant position is a direct consequence of the success of the logicism of Frege and Russell in logic at the outset of the twentieth century. Second, there are precedents for using an alternative model in the interpretation of Genevan findings, notably by Piaget and Garcia (1987/1991) who explicitly set out to "clean up" Piaget's logic using an alethic model. It is also interesting that Piéraut-Le Bonniec (1980) took a similar stance, even though her own position made reference to an alethic modal logic applied to action. This opens the door to the use of deontic logic instead. Third, it is evident that some of Piaget's own findings are unwitting testimony to children's deontic reasoning, but these findings were not expressly interpreted in his analyses (Smith, 2000b). In general, there is a major lacuna here in current research on "pragmatic reasoning schemas" as reviewed by Fiddick *et al.* (2000). This research has a bias in favour of the late onset of reasoning about conditionals, and much less the development of practical reasoning based on directives charged with deontic modality.

Chapter 8

Educational Implications in a Constructivist Model of Education

The main aim in this chapter is to identify educational implications of the main argument in the previous chapters that reasoning by mathematical induction is under development during childhood. Since this argument had its basis in Piaget's developmental epistemology, the discussion of its educational implications will make the same commitment. To that end, the discussion is in two parts. One is a general review of Piaget's constructivist model of education. The other is a specific review of the educational implications of the present argument.

At the outset, an omnibus objection should be discussed head-on. The objection is widely accepted and amounts to what is currently the default view of Piaget's educational model. This default view states that Piaget's model of education is completely flawed and so should be pensioned off. But this objection is itself open to the counter-objection that it is an *ignoratio elenchi* both in imputing an interpretation which does not square with Piaget's own position and in by-passing alternative interpretations (Smith, 2002c,d). Interestingly, this is not a novel phenomenon (Parrat-Dayan, 1993).

The invalid objection has a starting point which is explicit in a famous paper due to Piaget (1970/1983, pp. 118–119):

> if a child, when he is counting pebbles, happens to put them in a row and to make the astonishing discovery that when he counts them from the right to the left he finds the same number as when he counts from the left to the right, and again when he puts them in a circle, etc., he has thus discovered experimentally that the sum is independent of order.

In this passage, Piaget has presented an example, even a paradigm case. True enough. But a case of what? Long ago, Socrates pointed out to his interlocutors that it is one thing to give an example; it is something else to state its underlying principle. To repeat, then: Piaget has presented a case, an example — yes; but an example of what? It is the interpretation — and specifically the interpretation stated or implied by the default view — which causes the problem. The preferred interpretation is false. Moreover, the presence of other interpretations never seems to be appreciated by the sponsors of the default view.

Under the default view, the interpretation of Piaget's example of the pebble-counting child leads to the following principles:

(i) The child is young and yet makes a developmental advance No special mechanism is set out. This means that the mechanism is due to one of Vygotsky's (1994) "two lines". One line is socio-cultural interaction (Davydov, 1995). The other line is biological maturation (Dehaene, 1997).

(ii)   The child is alone with neither peers, nor parents, nor teachers in sight to provide the necessary social interaction and guidance in the construction of knowledge (Holzman, 1997; Mercer, 1995).

(iii)   The learning context is physical in that the child is counting stones. As such, this context is impoverished and devoid of cultural tools and generational legacies typically available in human societies (Bruner, 1996; Kozulin, 1998).

(iv)   The child's discovery is abstract logical knowledge whose fit with reality is hard to fathom (Smedslund, 1966). This means that its fit with educational practice is problematic (Galton, 1999; Whitburn, 2000).

Using these principles, the objection amounts to a disjunctive syllogism. There are two possible mechanisms in (i). But in Piaget's model of education, one of these (socio-cultural interaction) is ruled out in (ii)–(iv). So this means that the other mechanism in (i) will have to be invoked. But this mechanism is biological maturation. Yet in education, biological maturation is a "distal factor" beyond the educator's control, unlike any "proximal factor" which is open to change in instructional interventions (Feuerstein, 1980). Since the case presented by Piaget does not identify any "proximal factor", this means that his model of education is flawed. *Ergo* it should be pensioned off — why would anyone make a commitment to a flawed model?

    This objection runs straight into a counter-objection. Premise (i) is false through the statement of a dichotomy in place of a trichotomy. What is ignored in a *tertium quid* or third alternative to Vygotsky's "two lines". According to Piaget, there is such a third alternative which he reckoned is equilibration in a constructivist model (for a critical discussion, see Smith, 2002a,b). Premises (ii) – (iv) are false. This will be demonstrated in the next section. This invalidates any inference through a disjunctive syllogism to the conclusion stated. Needless to say, the default view also by-passes two recent texts with an educational relevance (Piaget, 1977/1995, 1998). In this context, the fallacy of hasty generalization is committed by "writing off" Piaget's actual model prematurely. Even commentators as acute as Bringuier (1977/1980, p. 128) have fallen into the trap of crediting Piaget with (educational) positions to which he never made any commitment, for example:

Bringuier	You formulated principles for restructuring the teaching of mathematics.
Piaget	No, no . . .
Bringuier	And didn't it result in a teaching method?
Piaget	No.
Bringuier	Oh, I thought it did . . .

This dialogue is an instructive reminder that there is more to Piaget's model of education than what is implied by the default view.

    In short, the counter-objection is that the default view is a caricature based on its grotesque interpretation. Piaget's model of education specifically requires there to be a pedagogical contribution from teaching in instructional settings. If only the sponsors of the default view had read more widely and wisely from elsewhere in the same paper:

> each time one prematurely teaches a child something he could have discovered for himself, that child is kept from inventing it and consequently from understanding it completely. This obviously does not mean the teacher should not devise experimental situations to facilitate the pupil's invention. (Piaget, 1970/1983, p. 113)

Many commentators (see Smith, 1993, p. 127 and Smith, 2002c) have seized on the claim in the first sentence about the self-construction of knowledge which is interpreted as (ii) above. They have done so without reading — or at least reading it in the context of — in the second sentence the *denial that this precludes teaching*. This undermines (iii) and (iv) above. What is opposed in Piaget's model is barren teaching which impedes intellectual development, a point that he made long ago in his commentary on Vygotsky's work (Piaget, 1962/1995, p. 333). What is endorsed in Piaget's model is teaching defined in terms of epistemologically sound principles. It is these principles which are set out in the next section.

8.1 Piaget's Constructivist Model of Education

Piaget's model of education includes a definition of education, educational aims, an epistemological rationale for education, principles of instructional assessment, and principles of instructional intervention. What follows is a review of each of these in the context of Piaget's constructivist model of education.

8.1.1 Definition of Education

According to Piaget, education is a two-termed relation linking

> on the one hand the growing individual [and] on the other hand the social, intellectual and moral values into which the educator is charged with initiating that individual. (Piaget, 1969/1971, p. 137)

Under this definition, education is a process of change in which the values invoked by the educator are made accessible to the learner. It follows that, under this definition, education has a normative element. This is because values have normative properties in that a value identifies something which is valuable, i.e. something is preferred over something else and in that sense is "a good thing", maybe even "better" than alternatives, or even the "best" there is (von Wright, 1983b). Since Piaget's definition makes values, and so evaluation, a constitutive (defining) property of education, it has three advantageous consequences.

First, this definition fits the proposal that education and development are two sides of the same coin (Case, 1985; Kohlberg, 1987). Development is regarded as change leading to improvement, to something higher (better) in standard and level, in contrast to regression which is to something lower (worse) in standard and level (Beilin, 1992; Moshman, 1998). As such, this is a value-laden definition of development. But if education is also defined through values, it too is value-laden. Therefore, either could be the counterpart of the other. The implication is that if education is about change in growing individuals which leads to the acquisition of knowledge, the crucial distinction is in terms of the growth of *better* — and not merely *more* — knowledge. Education leading children to acquire more knowledge is not enough just because an incremental increase in knowledge does not in itself amount to improvement, i.e. to something which is better than what was there before.

Second, "Hume's rule" which states that values are never derivable from facts alone is widely accepted (von Wright, 1983a). Piaget's definition is compatible with "Hume's rule" since it does not require any individual to derive his or her values from facts alone. This is because the acquisition of any knowledge is never solely cognitive, but instead always has an affective and axiological component (Brown, 2001). Indeed, this is a central principle in Piaget's (1977/1995) social account, notably in his model of value-exchange where individual development could not be completed without socio-cultural interaction (Kitchener, 1991; Mays, 2000; Mays & Smith, 2001).

Third, Dewey (1966) argued that just as the geometrical definition of a straight line is in terms of the shortest distance between two points, so the definition of education required the identification of "two points". Dewey regarded these to be the child and the curriculum. One of these (the growing child) is common ground between Dewey and Piaget. They differ as to the other. For Dewey, a curriculum would identify what to teach and how this should be taught. For Piaget, values would identify why something should be taught or learned in the first place. This means that Piaget's definition operates at a more basic level than Dewey's in virtue of its capacity to explain why a specified teaching method or learning procedure is worthwhile (i.e. valuable) at all.

Piaget's definition of education does not set out "the" values of education. Instead, it covers the values invoked by the educator. This means that this definition fits value diversity which is a fact of social life. Thus training in the Boy Scouts (Piaget, 1998, p. 50), and political induction into the Hitler Youth (Piaget, 1977/1995, p. 25) are both covered by the definition, even though these are different from each other as well as from learning in formal settings such school subjects as history, science, art, and mathematics (Piaget, 1998, pp. 97, 177, 199, 231). In short, Piaget's definition of education is comprehensive. But in itself it is not enough since something more needs to be said about the identification of values, as too principles to their instructional function.

8.1.2 Educational Aims with an Epistemological Rationale

In Piaget's model, the central aim of education is as follows:

> education, for most people, means trying to lead the child to resemble the typical adult of his society [whereas] for me, education means

making creators, even if there aren't many of them, even if one's crea-
tions are limited by comparison with those others. (Bringuier, 1977/
1980, p. 132)

This statement can be explained through the distinction between the transmission and
transformation of knowledge. The transmission of knowledge in normal societies is a
necessary part of education. This was explicitly accepted by Piaget (1977/1995, p. 57),
notably in virtue of the inestimable socio-cultural contribution made by previous
generations to the growth of children's minds. But a necessary condition is not a sufficient condition. If it were, how could there be any advance since available knowledge
would endlessly perpetuate itself through the generations. Taking such a stance to its
logical conclusion was an Esquimo who told an anthropologist that "we keep to our
old customs so as to preserve [*conserve*] the universe" (Piaget, 1998, p. 111). The
transmission of available knowledge has substantial benefits in that

Ideas which have been painfully "invented" by the greatest geniuses
become, not merely accessible, but even easy and obvious, to school-
children. (Piaget, 1977/1995, p. 37)

In other words, the transmission of knowledge is not merely compatible with Piaget's
model of education, but is required by it. But transmission is not enough since there
is the transformation of knowledge to take into account as well. This is the second
aspect of Piaget's statement of his main aim of education. The transformation of knowledge is the development of new knowledge. Although this can take place in an
indefinite number of ways, two of these stand out.

One is where knowledge is novel at the level of common culture and society, i.e.
novel for anyone at all. Einstein's theory of relativity is just such a case. Another was
quantification theory in logic (Frege, 1881/1979; see Kneale & Kneale, 1962). Novel
knowledge in this sense is a real "first", comparable to Neil Armstrong's first human
footstep on the Moon. Knowledge may be novel in a different way, when it is novel
at the individual level, i.e. novel for any one person. The central limit theorem was
proved independently by Turing, even though his belief that this was the first demonstration turned out to be false (Hodges, 1983). Less illustrious cases are legion and
were noted in the previous paragraph as intellectual acceleration during schooling.
Indeed, instructional acceleration is the central principle in the CASE — Cognitive
Acceleration through Science Education — interventions (Adey & Shayer, 1994;
Shayer, 1999). An inventive mind can generate an individually novel way of dealing
with a standard problem. One example was given in Chapter 6 in Box 6.2. Here is
another from a five year old child taken from the same second study (*equals were
added to unequals*) where one of the containers ("greens") was given a "head start"
over the other ("orange") after which an equal number of counters was then serially
added. The dialogue starts just after Mary had agreed that there are more green cats
than orange cats in the two containers (see Box 8.1). The inference is correct, namely
that in this context there *cannot* (a modal claim) be an equality, from which it follows
that its opposite is the case. That is, there is an inequality. *QED*. But there is more

---

Box 8.1: Mary's reasoning about what "has to be"

Interviewer   If the Greens got a head start and there is more in the Greens, there
                has to be more in the Greens, does there, or not?
Mary*   Yes.
Interviewer   There does — why?
Mary   Because in a running race if someone, if a green cat and another
                (an orange cat) have a race, and the green sneaks off and runs its
                fastest, then the green wins. The other cat comes in at the end. So
                if the Greens get a head start and the Oranges don't, then there
                can't be the same in each.

* not the child's actual name

---

here than true knowledge. This child was child 88, and so many of the other children
had reached the same (correct) conclusion beforehand. All the same, none of these
children had expressed their knowledge in this way, based on an analogy of running
a race where the winner "sneaked off" at the outset and then ran at full speed.
Throughout this contribution, Mary smiled politely, but ironically, in accompanying
what was said with a manual demonstration of the main point in the analogy whereby
two fingers on one hand sped across the table whilst two fingers from the other trailed
behind. Evidently it was Mary's view that the interviewer was in need of assistance
by asking such obvious questions with an air of mystification. The analogy was offered
to put the interviewer's mind at rest. Here, then, is a case of knowledge where individual differences in the acquisition and display of knowledge are salient. This was
novel knowledge put to use by this individual child.

One consequence of this educational aim is that education should be tied not to a
content-curriculum but rather to a thinking-curriculum. In a paper given at a conference at the end of the Second World War, this difference was expressed by Piaget
(1998, p. 163) as the prescription that "it is necessary to teach children to think". There
are three points here. One is that this is an instructional aim about teaching. At a stroke,
this refutes the default view according to which Piagetian pedagogy is teaching-free.
The second is that teaching always requires something to be taught, and the proposal
here is that this is thinking. As such, this claim predates and is in line with recent
educational proposals about teaching directed on "reasoning as the 4th R of schooling"
(Nickerson, 1985; Resnick, 1987) or "the thinking curriculum" (Coles, 1993; Leat,
1999). The third point is that this is a requirement, not an option. The implication is
that children left to their own devices will not automatically learn to think since
other intellectual capacities are equally natural — for example, mindless copying and
idleness — a point made long ago, in 1935, by Piaget (1969/1971, pp. 137–138; 1998,
p. 155).

8.1.3 *Epistemological Rationale for Education*

The epistemological framework set out in Chapter 7 is applicable to Piaget's educational model. Central to this framework are five parallels in the work of Frege and Piaget under the framework AEIOU (autonomy, entailment/equality, intersubjectivity, objectivity, universality). An example of this framework was given in Box 6.2. Further illustration and discussion of this framework is available elsewhere (Smith, 1999b,c, 2000a, 2002a,b).

The relevance of this framework to education is due to the importance assigned to teaching as the bringing about of learning which is the acquisition of knowledge. This is where the unit of analysis proposed in Chapter 7 fits in. The acquisition of knowledge in a constructivist perspective is an act of judgment. Any such act has causes, such as the teaching which led to this learning. But if the act is also an act of judgment, there are normative issues in the reckoning as well. The AEIOU framework identifies five of these normative properties. Under this proposal, new knowledge is based on a process which is:

> *autonomous*: displayed as spontaneous thought whose grounds are independent of the causal conditions generative of that thought.

In school learning, Piaget's (1969/1971, p. 222) proposal is that children should not do what they want; rather, they should want to do what they do. A learner should be a free agent, but should not have free licence (Piaget, 1998, pp. 213, 259). Autonomy is not anarchy since rational thought and self-indulgent thinking are not the same thing (p. 165). Epistemological freedom is intellectual liberty subject only to normative constraints which are intrinsic to the knowledge under construction. Learners are no more free to suspend the law of gravity than they are to suspend the law of contradiction. Yet in anarchic learning, "anything goes", including the law of contradiction. Autonomy also excludes heteronomy, manifest as conformity to the intellectual norms imposed by any teacher. Social conformity may well occur as an act of heteronomous obedience in a prevailing culture where the teacher becomes "the symbol of knowledge and ready-made Truth, of intellectual authority and Ancient tradition" (p. 162). Obedience due to the "the halo of explicit or implicit authority which is attached to the master's word or to the textbook" is not enough (p. 216). Children may be changed by their schooling but not in an epistemologically valuable way. It is one thing for reasoning to be an act of obedience, i.e. one individual accepts something on the basis of another's authority. It is something else for obedience to be an act of reason, i.e. an individual accepts something in virtue of the laws internal to a knowledge-domain (Piaget, 1977/1995, p. 60).

> *entailment/equality* (modal knowledge of necessity): displayed as understanding in virtue of necessary rather than contingent laws and so based on "what has to be".

Children may learn to reason correctly either by using mathematical principles (Gelman & Williams, 1998), or by making logical inferences (Nunes & Bryant, 1996).

Both are important advances. But learning to reason modally is another matter, concerning "what had to be so, and could not be otherwise". Such reasoning is based on entailment along with equality which are logically necessary relations (Smith, 1997, 1999d). All logical deductions and all mathematical truths are necessities. Normativity entails necessity, which is no doubt the reason why necessity was regarded by Kant (1933) as one criterion of good understanding. This is because intellectual development is an interplay between the normativity of thought and the subjectivity of thinking. "Before reasoning in the normatively right manner, the small child begins by playing with his or her thought or by using it *in accordance with norms which are unique to that individual*. Thus the principal task for the education of the intellect, it seems, is much more to form thought and not to furnish memory" (Piaget, 1998, p. 148; emphasis added). A norm lays down "what has to be" or "what must be done". Personal modes of thinking amount to "what has to be for me". Deductive necessity occurs in "correct reasoning on the basis of a hypothesis which is recognized as false, in saying: if that is admitted, then it follows *necessarily* . . ." (p. 248; emphasis added). Such reasoning requires a distinction between relations which are logically necessary, and relations which are not. Necessities are not pseudo-necessities. They are not "false absolutes" such as Newton's "absolute space and time" (p. 92). Nor are they cultural norms such as those attributed to the Esquimo above and in commenting on which Piaget added the telling remark that "from a social point of view, we are all like this primitive man" (p. 111).

> *intersubjective*: displayed in the recognition that one and the same thought can be grasped by different individuals.

Piaget agreed with Frege (1918/1977) about the importance of self-identical thought. A thought such as the Pythagorean theorem is a self-identical thought just in case it is one and the same thought which is represented either on different occasions in one mind or on any occasion in different minds. The intersubjectivity of thought is acknowledged in the claim that in logic "proof is valid for all" (Piaget, 1924/1928, p. 24). A logical structure is just such an instrument in that "logical structures constitute the sole instrument common to demonstration used in every science" (Piaget, 1949, p. 2). Intersubjectivity is secured by Piaget's epistemic subject. This subject is defined through the use of a logical structure which embodies "a general logic, both collective and individual, which characterizes the form of equilibrium common to both social and individual actions" (Piaget, 1977/1995, p. 94). In this same sense, intersubjectivity is invoked in Piaget's (1998, pp. 37, 38) reference to "mutual agreement . . . mutual respect". Education should be directed on the "formation of a social attitude of reciprocity capable of being generalized in progressive steps" (p. 260). Reciprocity in thinking is a *sine qua non* of the intersubjectivity of thought. As such, it is required by rational agreement and disagreement. Without some common ground, there could not be a meeting of minds at all.

> *objective*: displayed in an acknowledgement that something is true independently of its being grasped.

Knowledge is objective just in case it is true. But it is one thing to make a response which is true. It is something else to acknowledge that the response is true. An act of judgment amounts to objective knowledge and so requires a recognition or realization of the truth of what is known. The use of a logical structure secures this acknowledgement in that formal logic is "the codification of the rules of truth" (Piaget, 1949, p. 4). There is a fundamental distinction between activity confined to "success or practical adaptation and verbal or conceptual thought whose function is to know and state truths" Piaget (1937/1954, p. 360). Objectivity requires both knowing what is true and that it is true (Piaget 1977/1995, p. 184; 1998, p. 115). According to Piaget (1998), this requires two conditions to be met. One is a "synthesis between expressions of the ego and submission to reality" (p. 201) and the other is "cooperation between individuals [in] a mutual critique and progressive objectivity" (p. 127). Both conditions turn on the view which was accepted by both Frege (1897/1979, p. 128) and Piaget (1949, p. 4; 1967/1971, p. 35) that logic is the science of truth. Judgments with a logical justification are in line with the laws of truth.

> *universal*: displayed as making a universalization, i.e. understanding what is true in all cases, even if this understanding is confined to one particular context.

Something may be universal in two different senses. One sense concerns the *transfer* of knowledge across tasks and contexts (Feldman, 1995). Another sense concerns the *universalization* of thought, i.e. understanding what is true of all cases. Empirical critiques of Piaget's work have centred on the former. But this critique is independent of the different question about the universalization of thought. This problem was explicitly stated in his first book, *Recherche* (Piaget, 1918, p. 46; see Smith, 2002a,b). A paradigm example is mathematical universalization, for example in reasoning by mathematical induction. As Piaget (1952) pointedly remarked, "the problem of universals [recurs] at all levels". Piagetian structures bear on universalization rather than transfer. The universalization of thought is the essential task of pedagogy which is "to lead the child from the individual to the universal [where] this ascent from the individual to the universal corresponds to the very processes of the child's intellectual and moral development" (Piaget, 1998, p. 81). Quite simply, "logic is a morality of thought just as morality is a logic of action" (p. 73; cf. 1932/1932, p. 404). On the basis of Chapter 2 above, a paradigm example of universalization is an inference on the basis of mathematical induction from what is true of a specified number to what is true about any number at all. On the basis of Chapter 6, this important capacity is under development during childhood.

To see why this AEIOU framework is appropriate, recall that Piaget's definition of education was in terms of the inculcation into a set of values. Now this leads to a problem about the formation of real values. The problem is not so much the formation of "our" values rather than those of others, but instead the formation of any value rather than a pseudo-value. The difficulty here is comparable to that of demarcating pseudo-concepts from real concepts (Vygotsky, 1994). Crucially, the problem concerns the reconciliation of *son moi et al loi*, the self and normative law (Piaget, 1977/1995, p. 241). This problem is general. Rules and practices in all domains and contexts have normative properties about "what has to be" and "what should be done". This is where the "loi" comes in. But these rules are understood by human minds in practices which depend upon human participation. And this is where the "moi" comes in. The problem which Piaget saw here is the problem of mismatch between the dictates of the former and the expectations of the latter. And this is a major problem because it arises for any model which regards the transmission of knowledge as essential in education. Exactly this commitment was made by Piaget (1997/1995, p. 76):

> each individual is led to think and re-think the system of collective notions.

Reconstructing available knowledge in its educational transmission is a condition of its reconstruction into new and better knowledge. Causally speaking, this is not a solitary matter since teachers and peers give social guidance, and are so accepted in Piaget's model. Normatively speaking, this is another matter. This is because the process of construction and reconstruction requires a continual reconciliation of *son moi et la loi*, between one's own representation and the requirements of reality. This problem is compounded by the fact that children live in social worlds which are responsible for a related problem of reconciliation between what can be regarded as *la foi et la loi*, ideological faith and normative law (adapting Piaget, 1918, p. 21; see also Smith, 2002b).

To clinch this point, consider a thought experiment which was analysed independently by Vygotsky and Piaget. Imagine a society of exact contemporaries whose members are the same age (for example, children aged seven years). This society has neither a common culture nor the benefits from the past. What would intellectual development be like in such a society? Vygotsky (1994, p. 351) ruled out intellectual progress due to the absence of cultural tools and more advanced members to guide social participation. Piaget (1977/1995, pp. 57, 289) agreed that the absence of transmitted knowledge would be a devastating loss. But this would not rule out development altogether. Although such a society would be disadvantageous, it would have compensations. One is the elimination of adult constraint (p. 149). Social rules are typically interpreted as directives which lay down "this is how we do it, so this is how it ought to be done". This problem has been characterized by von Wright (1983a) as the problem of "normative pressure". This is because values have normative properties which set out what should, or may, or ought not to be done (von Wright, 1983). As such, values make a determining contribution in the direction of action (Ross, 1968). But it is one thing to accept someone else's value on the basis of their authority. This amounts to heteronomy. This is likely to happen in the transmission of cultural values which are present from the cradle to the grave (Piaget, 1977/1995, pp. 278, 286). It something else again to accept something through one's own autonomous account in an intersubjective, objective and universal system of thought. Note that these properties are secured in the AEIOU framework. Normative pressure would be reduced — but not eliminated — for the children in the thought experiment, who would still have the task of devising their own social rules, and then creating better ones. The main point behind Piaget's analysis of this thought-experiment is that in a causally non-standard society the situation which is faced by the growing individual is norma-tively the same problem.

The relevance of this to Piaget's model of education is that teaching children to think is the central task of education. But this is notoriously not an easy matter. One reason why this is difficult is in view of the normative properties of knowledge, reasoning and "good thinking". Some of these normative properties are identified in the AEIOU framework. This quintet is not presented as a complete list. To the contrary, it is massively incomplete (see Brown, 2001; Demetriou & Kazi, 2001; Donaldson, 1992). But it does secure one main point about development and education. Under this proposal, a unit of analysis is an act of judgment (Smith, 2002a,b). Any act is an action which has causal antecedents for investigation in psychology. Any judgment has an implicatory content for investigation in (developmental, empirical) epistemology. The AEIOU framework marks some of the properties of this implicatory content. An adequate account would cover both. Yet most accounts in both psychology and educa-tion cover one of these ("causal facts") rather than the other ("normative facts"). This is a paradigm case of what Piaget (1977/1995, p. 51) identified as one of the major aspects of developmental advance which is "this transition from the causal to the logical".

8.1.4 Principles of Instructional Assessment

Assessment in education occurs in multiple forms, including diagnostic, formative and summative assessment (Black, 1998b; Goldstein, 1991). The interpretation of any assessment measure occurs in two ways, either as norm-referencing or as criterion-referencing (Black, 1998a; Satterly, 1989). The main focus in Piaget's model is diagnostic and formative assessment interpreted through criterion-referencing.

The educational focus on diagnostic assessment has long been recognized as one essential first step in all teaching. In their major review, Black and Wiliam (1998) have argued that the importance of this type of assessment is currently inversely related to its successful use in classrooms. Its importance is not in doubt. Following Aristotle (1987, sect. I.1) in his *Posterior Analytics*, "all intellectual learning comes from already existing knowledge". Ascertaining prior knowledge is therefore the first task in teach-ing and was so accepted by Piaget (1998, p. 26): "the primordial question is to know what are the child's available resources". This is a diagnostic question. The aim behind the assessment is to identify what is known, and what is not known, by any learner. Between these two poles lie all the intermediaries from minimal to maximal compe-tence (Flavell *et al.*, 1993). In Chapter 5, a review of Piaget's critical method was set out through its three main features, namely judgment as the unit of analysis, standard-ized activity for contextualization, and explication of the judgment's grounds. The key feature of this method is that it requires both responses and reasons which start from standardized task settings but which can be adapted to fit the profile of individual learn-ers. The use of such a method in both child psychology and education has been elab-orated by Ginsburg (1997). Although most of Piaget's own studies were cross-sectional, longitudinal designs were notably used in the infancy studies (Piaget, 1936/1953,

1937/1954). These studies may have been one basis for micro-genetic designs recommended by Inhelder *et al.* (1974; see Vonèche, 2001) and recently reviewed by Miller and Coyle (1999).

The educational focus on formative assessment is also evident in Piaget's model. There are two reasons why. One is that Piaget's (1960) stage-criteria include a bi-directional requirement whereby learning at any level is related both to its precursors and to its successors. In a developmental epistemology,

> all knowledge can be considered as being relative to a given previous state of lesser knowledge and also as being capable of constituting just such a previous state in relation to some more advanced knowledge. (Piaget, 1950, p. 13)

Inhelder's (1954/2001) tasks directed on children's thought are exemplary in this regard in leading from practical intelligence in early childhood to formal thought in adolescence. The second reason is Piaget's *tertium quid*, or third alternative in equilibration, which locates internal construction as an interdependent factor in developmental advance. The functioning of internal construction was interpreted in Chapter 7 in terms of the interplay between making a learner's response and the reasons given by that learner. These reasons typically involve interlocutors and teachers have a role to play here. Piaget (1973, p. 85) regarded the teacher to be less "a person who gives 'lessons' and is rather someone who organises situations that will give rise to curiosity and solution-seeking in the child, and who will support such behaviour by means of appropriate arrangements". This is because "the role of *the teacher becomes central* as the animator of discussions *in consequence of having been the instigator*, within each child, of the taking of possession of that remarkable power of intellectual construction which is manifest in all genuine activity" (Piaget, 1998, p. 191; emphasis added). It is evident that formative assessment in Piaget's model amounts to dynamic assessment (Feuerstein, 1979). Although dynamic assessment is often interpreted through Vygotsky's (1994) model applied to education (Shephard, 2000; Torrance & Pryor, 1998), there is no requirement for this in view of the availability of Piaget's model.

The educational focus on criterion-referencing is evident in the British National Curriculum (DES, 1991; DfEE, 1999). According to Black and Wiliam (1998; see also Black, 1998a), the tendency in education is for ostensible criterion-referencing to become norm-referencing, or in other words for assessment in terms of a designated standard to be converted into assessment in terms of peer-comparison. Indeed, a spectacular form of this is international testing in mathematics (Reynolds & Farrell, 1996; Whitburn, 2000). It is evident that criterion-referencing, and not norm-referencing, is required by Piaget's model. The reason is clear. His model is an epistemological model. As such, it is directed on levels in the development of true knowledge. But learning and the development of knowledge are central to school learning. This is explicit in the British National Curriculum:

> the National Curriculum must promote development in all main areas of learning and experience. (DES, 1989; see also DfEE, 1999)

Evidence of reliable scaling based on Piagetian levels can be found in both national (Bond, 1997, 2001; Shayer & Adey, 1981) and international (Shayer *et al.*, 1988) studies. This means that these levels could provide a basis for criterion-referencing in education.

8.1.5 Principles of Instructional Intervention

The first point to make about Piaget's pedagogy is that there is one. This is a point worth making just because this is denied by many sponsors of the default view set out in the previous section. Bruner (1960) famously set out an ambitious proposal that any child could learn anything in any domain in some intellectually honest form. This proposal was widely regarded as incompatible with a Piagetian model which faced a dilemma. Either Piaget's model had no room for teaching at all (Ginsburg, 1981). Or Piaget's model of teaching trailed behind developmental advances made by learners in view of its overcommitment to readiness (Kuhn, 1979). But this dilemma is false since it arises only under an interpretation of Piaget's model. It is pretty clear that the interpretation on offer here is incompatible with both alternatives.

It is incompatible with the first alternative since teaching is defined in, and therefore not excluded from, Piaget's model.

> A teacher creates a learning context which evokes a spontaneous elaboration of the part of the learner. (Piaget, 1962/1995, p. 333)

This position is comparable to that held by Vygotsky (1994, p. 366):

> a teacher creates a learning context where higher concepts arise from lower concepts in the learner's mind but such that under no circumstances can higher concepts be deposited in the learner's consciousness from outside.

This is a telling point in that Vygotsky here denied that new concepts can be deposited into a learner's mind from outside. Yet any teacher is in just such a position in relation to any other learner and so is thereby excluded from "making" someone else understand. The reason for this is human autonomy which is interpreted here through the AEIOU framework as a key feature of Piaget's model. A teacher can of course secure heteronomous learning whereby children learn to give "correct answers". But this is not the same thing as thinking for oneself which is required by a commitment to "reasoning as the 4th R of schooling" (see section 8.1.2). According to Kamii (1982), much schooling is directed on the production of "correct answers". One of her examples was a boy who argued that $3/4 + 1/4 = 4/8 = 1$. This amounts to incompatible answers to a question based on the use of (heteronomously) acquired procedures: "you get a different answer every method you use" (p. 322). This view of the questionable quality of school learning is shared by others (Raven, 1994). At any event, Piaget's pedagogical point is that good teaching requires the design of good learning contexts. His point has been well taken by some (Adey *et al.*, 1989; Case, 1985), even though there is clearly scope for further research and development here.

The second alternative is also incompatible with the interpretation here. Some commentators such as Bruner (1966, p. 29) have interpreted Piaget's model as a commitment to readiness, from which it follows that teaching would lag behind developmental advance. This means that teaching is futile unless the child is ready to learn. This is reckoned to be bad enough, since it requires teachers to wait until children are ready to learn. Even worse, it denies a role to teaching altogether since children who are ready to learn do not require to be taught at all. True, it may seem that this recommendation has its basis in Piaget's (1998, pp. 194, 263) injunction for researchers and teachers to "follow the child's natural development". But this is not the case. First, this injunction in context concerns development or progress in learning and the recommendation is about the *study* of children so as to establish developmental laws. Second, teaching and learning are not the same thing (Desforges, 1995; Fenstermacher, 1986). *A fortiori* neither are teaching and development since the latter amounts to complex learning directed on progression. Third, laws of nature do not generate their applications. Applications in engineering are all in line with physical laws. But automobiles and computers did not spontaneously emerge of their own accord, nor did engineers simply wait until physical nature has taken its course. Rather, they engaged their creative expertise in due cognisance of physical laws. Similarly, teachers and educators should engage their creative expertise in due cognisance of developmental laws. There is a spectacular mismatch between the recommendation attributed to Piaget by his commentators, that teachers should await children's readiness, and Piaget's own pedagogical stance in the previous paragraph. It is autonomy which lies behind Piaget's remark quoted above that it is better to let a child discover a truth than to teach that truth ready-made. At issue is *not whether* to teach mathematical concepts, *but rather how* to teach so that what is taught is understood. Quite simply, not all teaching produces good learning.

> In some cases, what is transmitted through education is well assimilated by the child because it represents an extension of spontaneous construction with consequential acceleration in development. But in other cases, educational transmission intervenes too early or too late, or in a manner that precludes assimilation because of a mismatch with spontaneus construction. In that case, the child's development is impeded, or even deflected into barrenness, as so happens in the teaching of the exact science. (Piaget, 1962/1995, p. 333)

Notice two good points here. One is a specific admission by Piaget that teaching can be effective, an admission which is denied in some commentaries (Case, 1985; Ginsburg, 1981). The other is the use of the term *spontaneity* in the sense of autonomy. But this does not mean the absence of teaching, *pace* sponsors of the default view in the previous section. The claim is that the knowledge due to non-autonomous learning will be poorly understood — it will amount to surface learning (Marton & Booth, 1997). Fourth, teaching can produce short-term benefit which is a long-term loss.

> There is all the more damage to the human formation of individuals by imposing truths on them from without, even when these truths are self-evident and even when they are mathematical, in as much as they could have discovered them for themselves. They have been deprived of a *method of research* which would have been more useful for their lives than the corresponding knowledge . . . in the formation of the whole mind. (Piaget, 1998, p. 175; emphasis added)

Children who gain high scores on school tests of knowledge today do not thereby have the capacity to develop their knowledge tomorrow. It is pointless for an educator to transmit today ready-made knowledge for use tomorrow since "we do not know what the society of tomorrow will be" (p. 120). schooling restricted to teaching without understanding is dysfunctional (Raven, 1994).

Two specific principles of instruction were identified by Piaget. One was a strong commitment to group learning.

> The active school *necessarily presupposes collaboration* in work (in that) group work is *in principle* more "active" than purely individual work. (Piaget, 1998, pp. 45–46, 158; emphasis added)

This explicit commitment directly refutes the default view. A second instructional commitment was to self-government in classroom learning.

> The method of self-government consists in attributing to pupils a share in the responsibility for scholarly discipline. (p. 167)

Self-government covers meteacognitive learning but it is also a wider, polymorphic notion which assumes different forms (pp. 47, 122). Examples include "the simple organisation of work in common by the pupils themselves, responsibility for collective discipline, extra-mural organizations (scholarly societies, clubs, etc.)" (p. 273). Other examples are national and international activities (p. 262) in as much as "nothing teaches the humanity of judgment and true modesty so much as daily contact with equals exercising free speech and possessing a spirit of comradeship" (p. 136). Self-government covers "rediscovery by oneself" (p. 46) as well as being "a process of social education, aiming — like all of the others — to teach individuals how to escape from their egocentrism so as *to* collaborate between themselves and to submit to shared rules" (p. 128).

These prescriptions were not made without a realization of the importance of evidence. As its director, Piaget made clear that the International Bureau of Education was in the business of collecting data on school learning for "educational science" (Piaget, 1998, pp. 48, 144–146, 178, 181, 194, 228, 263). But the evidence on offer in this same text is, and is so stated by Piaget, to be underdetermining, since there is no specific evidence used as evidence. What is missing is evidence from schooling about a common "psychological instrument based (*fondé*) on reciprocity and cooperation" (p. 120). In this regard, the extensive project undertaken by Adey and

Shayer (1994; see also Adey *et al.*, 1989; Shayer, 1999) is some compensation since their work shows that a classroom intervention designed on Piagetian lines can "really raise standards" in terms of an educationally significant improvement in performance on national tests in three core subjects (English, Mathematics, Science) in the British National Curriculum.

An argument based on "division of labour" could be invoked at this point. Piaget did state that his problems were primarily epistemological (in Hall, 1970, p. 25), no doubt in recognisance of his own work (Piaget, 1950). Further, there is a theoretical compensation for empirical incompleteness. In Piaget's epistemology, *all knowledge* is regarded as intrinsically social.

> Human knowledge is *essentially* collective and social life *constitutes* an essential factor in the creation and growth of knowledge, both pre-scientific and scientific. (Piaget, 1977/1995, p. 30)

This is a strong claim whose implications are poorly understood in Piagetian commentary (Mays & Smith, 2001). It is by interacting with other individuals in groups that any learner is in a position to develop in an epistemologically beneficial manner. The intersubjectivity of knowledge is secured by one individual's reference to the knowledge of other individuals. The objectivity of knowledge "presupposes the coordination of points of view, which implies cooperation (which is) essentially the source of rules for thought" (Piaget, 1998, p. 151). Self-government requires "real autonomy in the classroom" (p. 167).

Piaget's model does, then, include pedagogical principles. Group learning and self-government are educationally important because autonomy, intersubjectivity and objectivity are epistemological values which are constructed in human exchanges with others. The education of the individual strictly requires engagements in human society and culture. It is interesting that — quite independently — recent perspectives in developmental psychology and education have assigned a special place to the social aspects of learning and to metacognition (Kuhn, 1997; Shayer, 1997).

In conclusion, there are three claims. One is that the default view, which is presented as a serious objection to Piaget's model, is open to counter-objection. The critique presented above shows that there are several interpretations of Piaget's model. Thus the question becomes which interpretation is the better. The second conclusion is that the distinctive elements in Piaget's model include a definition of education, educational aims, an epistemological rationale for education, principles of instructional assessment, and principles of instructional intervention. In each case, an exegetical review has been presented which is both at variance with the default view and in line with current positions in education. The third conclusion is that development is defined epistemologically in Piaget's model in terms of the questions which ought to be addressed in knowledge-domains directed on progress in knowledge in that domain. The pedagogical counterpart is this. Teaching is defined epistemologically in terms of reducing — though never eliminating — two types of mismatching, one between these questions and the answers in fact given to them by learners, and the other between these questions and coincident questions which learners in fact ask (Piaget, 1998, pp. 147, 191; cf. 1962/1995, p. 333).

8.2 Educational Implications in a Constructivist Pedagogy

The review of Piaget's model of education in the previous section provides a general framework in which can be located specific implications arising from the present study. Six implications for child development and education are now set out. These are: principled problems; Piaget's model; children's reasons; children's mathematical reasoning; psychology and education; and education.

8.2.1 Principled Problems

The empirical study in Chapter 6 had its basis in a neglected study from 1963. But the reason why the neglected study was reckoned to be important was not because it was due to Piaget and Inhelder, but rather because it dealt with a problem of intrinsic merit, namely mathematical induction. It turned out that this type of reasoning has been by-passed in research in developmental psychology and education applied to childhood. From a constructivist perspective, this is a significant omission just because constructivism is incompatible with novel capacities emerging from nothing. Piaget (1942) was a notable opponent of *ex nihil* positions, which have been styled by Siegler (1996) "immaculate conceptions". One implication of the present argument is that principled problems can function both as heuristics (search for principled problems amenable to empirical investigation) and as hypotheses (such as those set out in Chapter 6). No doubt mathematical induction is not the only case, but it does at least provide a fine example of a problem which has both exercised some of the best minds — Poincaré, Frege and Russell — and which can be recast into a form which is intelligible to children. A standard aim of education is to ensure that children in one generation have access to the best knowledge from previous generations (Case, 1985). A related aim is to give them the opportunity to rethink that knowledge (Piaget, 1977/1995). This means that the stance taken in this study could be generalized. Principled problems are usually paradigm cases of the best knowledge from the past and, as such, deserve to be rethought by members of each generation. It was pointed out in the previous section that this was regarded by Piaget as one main function of education.

8.2.2 Piaget's Model

Piaget's verdict on the reception of his work was that he had been the most criticized psychologist of all time, but had come out alive (quoted in Smith, 2002a). Beilin (1992) was nearer the mark in comparing Piaget's contribution to that of Shakespeare on English. Other developmentalists are not sure about their legacy as "heirs to the house that Jean built" (Scholnick, 1999). Lourenço and his colleagues (Lourenço & Machado, 1996; Machado *et al.*, 2000) have made out a case that Piaget's model has been interpreted without due regard to Piaget's own statements of position and to alternative interpretations. There is a weaker position which suffices here, namely that Piagetian ideas and evidence continue to be productive. Under the interpretation of Piaget's model set out in Chapters 7 and 8, the upshot is that there are two questions to confront, one about "causal facts" and another about "normative facts". Causal facts (what are the causal conditions of the display of such-and-such reasoning?) are central to developmental psychology and education. Normative facts (how are the normative criteria of such-and-such reasoning developed by agents in their reasoning?) are central to developmental epistemology. Both are important. Both are also empirical. Neither can be answered in terms of the other. This interpretation — rather than the default interpretation implicated at the outset of this chapter — is important with regard to the continuity between (even identity of) development and education. One way to act on this view is to consider and reconsider Piagetian ideas and evidence, including his constructivist pedagogy reviewed earlier in the present chapter. Learning and development in instructional settings are in line with this interpretation on two counts. First, teaching is defined through its causal effects on learning (Desforges, 1995). Second, progression in learning amounts to development in virtue of a normative advance to better learning (Kohlberg, 1987).

8.2.3 Children's Reasons

Children in School Year 1 and Year 2 (age range 5–7 years) were repeatedly asked to make responses and to give reasons for them. The evidence from Chapter 6 indicated that this phenomenon (the capacity to give reasons) is important in its own right. The importance of reasons, and not merely "correct answers", was noted in Chapter 5 to be well taken in educational research. The implications for educational practice are another matter. The conclusion drawn by Black and Wiliam (1998) in the previous section is a stark reminder that much is still to do with regard to both the initial training of teachers and also in-service professional development with regard to diagnostic assessment in the classroom. The implications for research in developmental psychology are another matter and so three points can be made. First, children's reasons are substantively important. Despite the recent interest in microgenetic studies (Miller & Coyle, 1999), a serious interest in children's reasons and the uses to which they are put is still something of a rarity. The common practice in developmental psychology is to reduce severely or even to eliminate these from the design and findings of empirical studies. Second, children's reasons are also important on methodological grounds through the argument in Chapter 5. That argument set out a case for collection and recording children's reasons, especially on reasoning tasks. Third, children's reasons are also important on epistemological grounds in line with the unit of analysis proposed in Chapter 7 in terms of acts of judgment. The oscillation between responses and reasons which function as retroactive organizations as well as proactive reorganizations is in line with Piaget's (1975/1985) model of equilibration (cf. Smith, 2002a,b). Although the capacity to give reasons has been central to the investigation of adults' argumentation (Kuhn, 1991), the present study shows that children's epistemology also merits attention. And this turns out to be a matter of some interest in education where very much work still remains to be done (Hofer & Pintrich, 1997).

8.2.4 Education and Reasoning

Although reasoning could well be regarded as the 4th R of schooling (Resnick, 1987; Leat, 1999), it tends to take second place in the mathematics classroom. In primary school mathematics, procedural skills of calculation and the principled knowledge of counting are given top priority, at least in British schooling (DfEE, 1997). This may be fair enough, but only if other equally important abilities are given due weight as well. Arithmetical reasoning is a case in point. In view of the importance of evidence-based theories — and by implication policies and practices — in Piaget's research programme, the educational use of his ideas and evidence could be expected to respect the advice that he gave on the last page of a famous study (Piaget, 1932/1932), and repeated elsewhere (Piaget, 1998; Bringuier, 1977/1980) namely that the educational implications of his work could *not* be simply "run off" from his developmental psychology and epistemology. A different way of putting this was given by Duckworth (1996); in moving from psychology to education, the wrong approach is to train children to complete Piagetian tasks in the mistaken belief that their intelligent understanding will be thereby improved, but a better approach is to encourage children, teachers and researchers to be inventive in their thinking. This is a salutary reminder of one way in which the present study could and should be augmented. At any event, if reasoning is educationally important, then modal reasoning is implicated. Two types of modal reasoning were noted in Chapter 7, namely alethic modality (necessity–possibility) and deontic modality (obligations–permissions). Neither type of modal reasoning normally figures on a school curriculum applied to childhood (cf. DfEE, 1999; DES, 1991). Yet both types of reasoning are relevant to a "thinking curriculum" (Coles, 1993). Almost everything is still to be done here in terms of educational design and evaluation.

8.2.5 Children's Mathematical Reasoning

The evidence presented in Chapter 6 was summarized at the outset of Chapter 7. A central conclusion was about the development of mathematical reasoning in children during school Year 1 and Year 2 (age range 5–7 years). This means that reasoning by mathematical induction was under development during this early schooling. There were constraints to heed as well, notably with regard both to the interaction between counting and reasoning, and to the demarcation of correct reasoning from modal reasoning. These findings would have to be replicated. They could also be extended to cover mathematical induction in older children as well as reasoning in areas of mathematics other than arithmetic. What is also clear is that this study was located in one knowledge-domain. This is in line with the position set out by Demetriou (1998;

Kargopoulos & Demetriou, 1998), though with consequential differences about the extent and functioning of domain-general processes. These include intellectual processes (Smith, 1998b). They would as well have to be restructured through affective and personality constructs intrinsic to the self (Brown, 2001; Donaldson, 1992; Demetriou & Kazi, 2001). The educational importance of such work could be considerable in view of their interplay in educational settings (Brown, 1997).

8.2.6 Psychology and Education

Piaget (1998) pointed out that his psychological studies had been put to a deplorable educational use, manifest as "teaching modern mathematics by archaic methods". He cited as a prime example teaching five year old children the notation of set theory or teaching number conservation. Apparently, not all commentators seem to have realized this (cf. Deheane, 1997; Hughes, 1986; Williams & Shuard, 1970). It would be a bad mistake to teach children to reason by mathematical induction using formal methods, whether these are "proof theoretic" or "model theoretic" (Johnson-Laird & Savary, 1999). Although reasoning by mathematical induction can be formalized, it does not follow that children should be taught to use formal methods in the first instance. The deplorable use of developmental psychology in education has been reviewed by DeVries (1987) in terms of the metaphor of translation from one language to another. Her central contention was that neither global nor literal translation are adequate, and so recourse to free translation is the sole alternative whereby the implications of a theory are used to drive applications to educational practice. There are two consequences worth noting. One is the undesirability of current practice in which mathematical reasoning is regarded solely as reasoning based on formal methods. This is overly restrictive since the findings from this study show that children in Year 1 and Year 2 reason by mathematical induction based on reasoning-in-action rather than on formal proof. The other is the undesirability of confining inductive reasoning in mathematics to adolescence for exactly the same reason. Both consequences apply to the British National Curriculum for Mathematics where mathematical proof is consigned to the highest levels of secondary schooling (DfEE, 1999). There is nothing in Piagetian constructivism to warrant this restriction. There is instead quite a lot in Piagetian constructivism to take forward the educational task of promoting reasoning throughout the whole of schooling.

Chapter 9

Logistic-normal Mixture Models Applied to Data on the Development of Children's Reasoning

Damon Berridge

In this chapter, the statistical analysis of Chapter 6 is taken a stage further through the development of a statistical model which will permit an assessment of the degree to which explanatory variables of interest such as year and study, having adjusted for explanatory variables such as age and gender, affect the chances of a child being able to perform a task. Such a model must be able to handle the following characteristics:

(i)   the binary nature of the responses (success/failure)
(ii) the possibility (indeed likelihood) that substantial variation between children will be due to unmeasured and potentially unmeasurable variables ("residual hetero-geneity").

9.1 Methodology

The ultimate goal of this section is to develop a model which will allow the multiple responses from a group of tasks performed by the children to be related to a set of explanatory variables, which will be a combination of categorical variables (gender, year and study) and a continuous covariate (age in months). However, for heuristic purposes this section begins with a much simpler scenario: a single binary response and a single binary explanatory variable in the first study. For the purposes of illustration, we consider Counting task Q1 (elastic non-stretched) in study I, and take year to be the single explanatory variable. An exploratory analysis of this scenario traditionally involves the construction of a table of frequencies cross-classified by the two variables of interest. This cross-tabulation is presented in Table 9.1.

A non-parametric test could be performed on Table 9.1. Several different types of non-parametric test are readily available. For example, SPSS for Windows performs a number of different tests. These range from the standard chi-squared test to the Mann-Whitney test which has been used in Chapter 6. However, the usefulness of these procedures for our purposes is severely limited. A non-parametric test can only

Table 9.1: Cross-tabulation of response to Counting task Q1 and Year.

	Correct response	**Incorrect response**	**Totals**
Year 1	35	15	50
Year 2	47	3	50
Totals	82	18	100

indicate whether there is a significant association between Q1 and year. An adequate explanation requires, in addition, an estimate of the difference in the chances of success between the children in Years 1 and 2, along with some measure of the precision with which that difference has been estimated. Simple tests like the chi-squared and Mann-Whitney can handle a small number of categorical variables, but soon become unmanageable with larger numbers of categorical variables. Furthermore, continuous covariates such as age generally have to be grouped into intervals, thereby losing potentially useful information. For these reasons, we are driven to abandon such simple non-parametric tests in favour of a statistical modelling approach, which will allow us to estimate not only the main effects of a wide range of explanatory variables (which could be categorical or continuous in nature), but also first- and higher-order interactions between such explanatory variables.

Table 9.1 can be used to calculate exactly the probabilities of a correct response from a Year 1 child and from a Year 2 child, which equal $35/50 = 0.70$ and $47/50 = 0.94$ respectively. Note that, in general, a probability can take any value between 0 and 1. These probabilities can then be used to compute the odds of a correct response from a Year 1 child and from a Year 2 child, which equal $0.70/(1 - 0.70) = 2.33$ and $0.94/(1 - 0.94) = 15.67$ respectively. Note that, in general, odds can take any value between 0 and ∞ (infinity). A measure of the difference in pattern of response between the two years is the odds ratio for the Year 2 children relative to the Year 1 children, which equals $0.94/0.70 = 1.34$. The odds for the Year 2 children may be expressed in terms of the odds for the Year 1 children and the (Year 2:Year 1) odds ratio as follows:

$$\text{Year 2 odds} = \text{Year 1 odds} \times (\text{Year 2:Year 1 odds ratio}) \tag{9.1}$$

Currently there are no safeguards to prevent model (9.1) from predicting a Year 2 odds between $-\infty$ and $+\infty$, outside the range of possible values $[0, \infty)$. A logarithmic transformation will provide that safeguard. Such a transformation of model (9.1) gives:

$$\log (\text{Year 2 odds}) = \log (\text{Year 1 odds}) + \log (\text{Year 2:Year 1 odds ratio}) \tag{9.2}$$

Let year be indexed by the variable y, with codes 0 (Year 1) and 1 (Year 2). Model (9.2) may be re-expressed in the following manner:

$$\log \text{odds } y = \log \text{odds in Year 1} + \log \text{odds ratio} \tag{9.3}$$

where the term "log odds ratio" equals zero for the Year 1 children ($y = 0$), and equals the (Year 2:Year 1) odds ratio for the Year 2 children ($y = 1$). Model (9.3) is called the binary logistic regression model (McCullagh & Nelder, 1989).

These log odds may be expressed in terms of p_y, the probability that a child in Year ($y + 1$) gives a correct response to Counting task Q1, $y = 0,1$. Model (9.3) may be written more formally as:

$$\text{logit}\{p_y\} = \alpha + (\beta \times y) \tag{9.4}$$

where

$$\text{logit}\{p_y\} = \log\{p_y/(1 - p_y)\} = \log \text{ odds for Year } (y + 1), \ y = 0, 1,$$

$$\alpha = \log \text{ odds for Year } 1 \ (y = 0),$$

$$\beta = (\text{Year 2:Year 1}) \log \text{ odds ratio}.$$

At first glance, this may appear to be a case of overkill. Why go to such trouble to analyse a single two-by-two table of counts? The major advantage of model (9.4) is that it may be readily generalized to handle much more complicated data structures. In order to utilize fully the richness of the experimental data introduced in Chapter 6, we can generalize model (9.4) in two distinct ways. These generalizations are discussed in subsections 9.1.1 and 9.1.2.

9.1.1 From One Explanatory Variable to a Set of Explanatory Variables

We wish to generalize model (9.4) to handle the two primary explanatory variables of interest: Year and Study, and the two variables of secondary interest: gender and age. Let us assume a child with vector of explanatory variables $\mathbf{x}$ responds to Counting task Q1. The probability of this child giving a correct response is denoted by $p(\mathbf{x})$. To handle the generalization from one explanatory variable to a set of explanatory variables, model (9.4) becomes:

$$\text{logit}\{p(\mathbf{x})\} = \alpha + \beta'\mathbf{x} \tag{9.5}$$

where

$$\text{logit}\{p(\mathbf{x})\} = \log\{p(\mathbf{x})/(1 - p(\mathbf{x}))\}$$

$$= \log \text{ odds of a correct response for a child with vector } \mathbf{x},$$

$$\alpha = \log \text{ odds of a correct response for a child with null vector } \mathbf{x} = \mathbf{0},$$

$$\beta'\mathbf{x} = \log \text{ odds ratio of a correct response for a child with vector } \mathbf{x} \text{ relative to a child with the null vector}.$$

The vector of regression coefficients β may be interpreted as the effect of the vector of explanatory variables on the pattern of response.

The binary logistic regression model may be fitted in most standard statistical software packages such as SPSS for Windows. However, for the reasons to be outlined in the next subsection, the present analysis utilizes the package SABRE (Software for the Analysis of Binary Recurrent Events) (Barry *et al.*, 1998). Further details of SABRE are available on the web:

http://www.cas.lancs.ac.uk/software/sabre3.1/sabre.html

The null binary logistic regression model

$$\text{logit}\{p(\mathbf{x})\} = \alpha$$

may be fitted in SABRE using the LFIT command as follows: LFIT INT. The explanatory variables may then be added to the model. For example, year (as represented by the SABRE factor FYEAR) may be added in the following way: LFIT INT FYEAR.

The deviance (or $-2 \times$ log-likelihood) of a model is a measure of the amount of variation left unexplained by that model. The change in deviance between the null model and the model with year included may be compared to the chi-squared distribution on one degree of freedom. If this change in deviance is significantly different from zero, then there is evidence of a significant year effect at the 5 percent level; in other words, there is a significant difference in the chances of success between Years 1 and 2.

The main effects of, and interactions between, the other explanatory variables may be added to the model in a similar manner.

9.1.2 From a Single Binary Response to a Repeated Binary Response

The binary logistic regression model (9.5) is sufficient to analyse responses to a single question such as Counting task Q1 in study I. Indeed, model (9.5) could be used to analyse responses to Counting task Q1 from both studies I and II simultaneously. However, the multiplicity of responses given by each child raises a number of modelling issues which are discussed in this subsection.

Model (9.5) would assume independence between responses to Counting task Q1 given by the same child on both occasions. Clearly, such an assumption is unrealistic. Responses from the same child are likely to be more highly correlated with each other than with responses given by any other child. There is a possibility (indeed likelihood) of substantial variation between children due to variables which may be unmeasured or even unmeasurable. A statistical model should handle such variation, commonly known as residual heterogeneity (for example, Davies & Pickles, 1985; Heckman, 1981; Massy *et al.*, 1970).

Study is indexed by the variable s, with codes 1 (study I) and 2 (study II). Let us assume that, in study s, child i with vector of explanatory variables $\mathbf{x}_{is}$ (which may now include a study indicator variable) responds to Counting task Q1, $s = 1, 2$. The probability of this child succeeding in this task on this occasion is denoted by $p_i(\mathbf{x}_{is})$,

$s = 1, 2$. In this chapter, we propose to handle residual heterogeneity by incorporating a child-specific random effect into model (9.5) in the following manner:

$$\text{logit}\{p_i(\mathbf{x}_{is})\} = \alpha + \beta'\mathbf{x}_{is} + \varepsilon_i \qquad (9.6)$$

where

$$
\begin{aligned}
\text{logit}\{p_i(\mathbf{x}_{is})\} \;=\; &\log\{p_i(\mathbf{x}_{is})/(1 - p_i(\mathbf{x}_{is}))\} \\
=\; &\log \text{ odds for child } i \text{ with vector } \mathbf{x}_{is} \text{ in study } s,
\end{aligned}
$$

α = log odds for child with null vector $\mathbf{x}_{is} = \mathbf{0}$ in study s,

$\beta'\mathbf{x}_{is}$ = log odds ratio for child with vector $\mathbf{x}_{is}$ relative to child with null vector,

s = 1, 2,

ε_i = random effect for child i.

The random effect is eliminated by integrating over the likelihood. Assuming a normal mixture distribution for the random effects, the likelihood integral can be evaluated numerically using Gaussian quadrature, and an estimate of the standard deviation of the normal distribution thereby obtained. The null logistic-normal mixture model

$$\text{logit}\{p_i(\mathbf{x}_{is})\} = \alpha + \varepsilon_i$$

may be fitted in SABRE using the FIT command as follows: FIT INT. The explanatory variables can then be added to the model, as described in the previous subsection. In addition, we can now check whether the effect of an explanatory variable remains constant from one study to another by fitting an interaction between that variable and the study indicator variable (declared as a SABRE "factor"). This model can also be used to analyse responses to more than one question simultaneously, for example, both Counting tasks Q1 and Q2.

9.2 Empirical Results

9.2.1 Counting Task (2 Questions)

The Mann-Whitney tests reported in section 6.2 revealed that the difference due to year was significant in study I for both questions Q1 (elastic non-stretched) and Q2 (elastic stretched), but that neither difference in study II was significant. This indicates the possibility that the year by study interaction may be significant in the logistic–normal mixture model of both questions simultaneously; in other words, that the (Year 2:Year 1) odds ratio may vary significantly from one study to the other.

The change in deviance (cid) upon adding age to the null logistic–normal mixture model is 9.3. Comparing this change in deviance to the chi-squared distribution on 1 degree of freedom (df) gives a p-value of $p < 0.01$; the age main effect is significant at the 1 percent level.

Having adjusted for age and gender, the study (cid = 61.7 on 1 df, $p < 0.001$) and question (cid = 70.9 on 1 df, $p < 0.001$) main effects are significant, as is the study by year interaction (cid = 4.22 on 1 df, $p < 0.05$), which means that the (Year 2:Year 1) odds ratio does indeed vary significantly from one study to the other. The (Year 2:Year 1) odds ratios for studies I and II are estimated to be 7.54 and 1.52 respectively. In study I, the odds of a Year 2 child giving a correct response are over 7 times greater than the odds of a Year 1 child responding correctly. By study II, this advantage to the Year 2 children had been reduced significantly. This confirms the convergence in performance of the two years that was reflected by the Mann-Whitney tests.

9.2.2 Conservation Task (3 Questions)

The majority of the children made correct responses to Q1 (initial selection) in both studies. Eleven children made one incorrect response, either in study I or in study II. A Mann-Whitney test performed on the pooled responses to Q1 in section 6.2 revealed no significant difference by year. An analysis of the pooled responses to Q3 (the modal question) in section 6.2 did not reveal a significant difference by year, nor were the separate analyses for study I and study II significant. However, there was a significant difference in responses to Q2 (correct reasoning) by year. But does this significant year effect remain once all three questions have been analysed simultaneously using the logistic-normal mixture model?

The standard error of the mixture distribution is not significantly different from zero, i.e. there is no evidence of significant residual heterogeneity between the children in their responses to these questions. Consequently, the logistic–normal mixture model reduces to the standard logistic regression model.

Having controlled for the significant age main effect (cid = 10.94 on 1 df, $p < 0.001$), both study (cid = 2.95 on 1 df, $p < 0.1$) and question (cid = 142.99 on 2 df, $p < 0.001$) main effects are significant. There is no evidence of a significant year effect in this task. Thus, having adjusted for age, the year effect, found to be significant in the Mann-Whitney test, is no longer significant.

9.2.3 Teams Task (2 Questions with 3 Subquestions/Versions in Each)

These questions generated ceiling and floor effects to Q1 (actual comparison) and Q2 (hypothetical comparison) respectively. Mann-Whitney tests outlined in section 6.2 revealed a significant difference in response pattern by year to neither Q1 nor Q2.

Question (cid = 259.5 on 1 df, $p < 0.001$) and version (cid = 6.93 on 2 df, $p < 0.05$) main effects are significant in the mixture model. However, there is no evidence of a significant difference in the pattern of response between the two years, thereby confirming the results of the Mann-Whitney tests.

9.2.4 Recurrence Task (2 Criteria)

This task was divided into four phases: A (initial state), B (actual and observed additions), C (actual and unobserved additions) and D (hypothetical additions).

Base criterion (phases B, C and D)   Mann-Whitney tests described in section 6.2 revealed a significant difference in the distribution of correct responses, for the six questions (two studies by three phases), between the two years. This indicates that the year main effect (and/or, possibly, interactions involving year) could be significant in the logistic-normal mixture model of all six questions analysed simultaneously.

Both gender (cid = 12.94 on 1 df, $p < 0.001$) and age (cid = 7.22 on 1 df, $p < 0.01$) main effects are significant in the mixture model. Having adjusted for gender and age, the year main effect is still significant (cid = 5.95 on 1 df, $p < 0.05$). The study (cid = 51.47 on 1 df, $p < 0.001$), and phase (cid = 18.79 on 2 df, $p < 0.001$) main effects are also significant in the presence of gender and age.

In the model with these significant effects, the (Year 2:Year 1) odds ratio is estimated to be 22.28. In other words, the odds of a Year 2 child giving a correct response are over 22 times greater than the odds of a Year 1 child responding correctly. The study by phase interaction is significant (cid = 8.71 on 2 df, $p < 0.05$). However, neither the study by year interaction, nor the phase by year interaction, is significant, implying that the above (Year 2:Year 1) odds ratio applies equally across both studies and all three phases.

Recursive criterion (phase D)   Mann-Whitney tests reported in section 6.2 revealed a significant difference in the distribution of correct responses, for the four questions (two studies by two questions: *great number, any number*), between the two years. This indicates that the year main effect (and/or, possibly, interactions involving year) could be significant in the logistic–normal mixture model of all four questions analysed together.

Both gender (cid = 16.59 on 1 df, $p < 0.001$) and age (cid = 22.94 on 1 df, $p < 0.001$) are significant main effects in the mixture model. Having controlled for gender and age, the study (cid = 10.71 on 1 df, $p < 0.01$) and question (cid = 61.99 on 1 df, $p < 0.001$) main effects are significant. However, the year main effect is not significant.

In the presence of the significant question and (albeit, not significant) year main effects, the year by question interaction turns out to be significant (cid = 11.7 on 1 df, $p < 0.001$). This result means that the (Year 2:Year 1) odds ratio varies significantly from one question to the other. The odds ratios for the questions *great number* and *any number* are estimated from the model to be 0.27 and 1.80 respectively. In other words, the Year 2 children performed significantly better than the Year 1 children on the *any number* question. However, the better performance on the *great number* question came from the Year 1 children.

Modal reasoning (phases B, C and D)   As with the base criterion, there were six questions (two studies by three phases), but in this case there was no indication of a significant difference between the years according to a Mann-Whitney test of the distributions presented in section 6.2. Even so, the overall incidence of correct modal responses was encouraging, amounting to just less than one half (46%) overall, with the correct modal responses given by 48 percent of the Year 2 children.

The modelling approach confirms the non-significant year effect that was observed in the Mann-Whitney test. Indeed, no significant effects at all could be found in the mixture model.

9.3 Empirical Discussion

The conclusions arising from the analyses in this chapter are broadly in line with the claims summarised and discussed in Chapter 7. Here we make specific reference to claims (i) to (vii) listed at the beginning of Chapter 7.

These analyses confirm the following claims:

(i)   A significant year effect in both versions of the Recurrence task in line with the base criterion.
(ii)   A significant year effect in the *any number* version of the Recurrence task in line with the recursive criterion.
(iii)   No significant year effect in both versions of the Recurrence task with regard to the modal criterion.
(vii)   A significant year effect on the Counting task, though the analyses have shown that the clear advantage held by the Year 2 children over the Year 1 children in study I had been reduced markedly by study II; no significant year effect on the Teams task.

However, the analyses do leave indeterminate two of the claims:

(ii)   The Year 2 children perform significantly better than the Year 1 children in the *great number* version of the Recurrence task in line with the recursive criterion; the analyses in this chapter indicate that, having controlled for significant gender and age effects, it is actually the Year 1 children who perform better in the *great number* version of the Recurrence task.
(vi)   Reasoning on the Conservation task was differentially interpreted; the analyses suggest that, having adjusted for a significant age effect, there is in fact no evidence of a significant year effect on the Conservation task.

The analyses in this chapter have not contained an epistemological element, so do not shed any further light on claims (iv) and (v).

9.4 Conclusions

The present analysis has achieved two interrelated objectives. It has applied statistical models for the analysis of repeated binary (i.e. success/failure) data in a major substantive area of contemporary educational research — the issue of reasoning by mathematical induction. Children drawn from Years 1 and 2 participated in up to five tasks in each of two cross-sectional studies. The statistical technique developed to model the effects of year and study on performance in these tasks incorporated two key statistical elements.

First, it allowed an element of control: we were able to assess the significance of the explanatory variables of primary interest (year and study), whilst adjusting for explanatory variables of secondary interest such as age and gender. Second, it modelled these binary responses within a modelling framework that took into account the problem of "residual heterogeneity" which is intrusive to repeated categorical data.

The results reveal a complex pattern with regards to the determinants of performance on the tasks. Year was shown to be a significant explanatory variable in the Counting task and the base criterion of the Recurrence task. We demonstrated that study was significant in all tasks, with the exception of the Teams task and modal reasoning in the Recurrence task. There was also evidence of a significant interaction between year and study in the performance of the children on the Counting task.

References

Adey, P., & Shayer, M. (1994). *Really Raising Standards*. London: Routledge.

Adey, P., Shayer, M., & Yates, C. (1989). *Thinking Science*. London: Macmillan.

Anderson, A., & Belnap, N. (1974). *Entailment: The Logic of Relevance and Necessity*. Princeton, NJ: Princeton University Press.

Apostel, L. (1982). The future of Piagetian logic. *Revue Internationale de Philosophie*, 142–143, 612–635. Reprinted in: Smith (1992), vol. 4.

Appelbaum, M., & McCall, R. (1983). Design and analysis in developmental psychology. In: P. Mussen (ed.), *Handbook of Child Psychology* (vol. 1). New York: Wiley.

Aristotle (1987). *A New Aristotle Reader*. Oxford: Blackwell.

Badia, P., Haber, A., & Runyon, R. (1970). *Research Problems in Psychology*. Reading, MA: Addison-Wesley.

Baltes, P., & Nesselroade, J. (1979). *Longitudinal Research in the Study of Behavioural Development*. New York: Academic Press.

Baroody, A. (1992). The development of preschoolers' counting skills and principles. In: J. Bideaud (ed.), *Pathways to Number*. Hillsdale, NJ: Erlbaum Associates.

Barry, J., Francis, B., Davies, R., & Stott, D. (1998). *SABRE: Software for the Analysis of Binary Recurrent Events: Version 3.1, A Guide for Users*. Lancaster: CAS Publications, Lancaster University.

Bechtel, W., & Abrahamsen, A. (1991). *Connectionism and the Mind: An Introduction to Parallel Processing in Networks*. Oxford: Blackwell.

Beilin, H. (1992). Piaget's enduring contribution to developmental psychology. *Developmental Psychology*, 28, 191–204.

Bell, V., & Johnson-Laird, P. (1998). A model of modal reasoning. *Cognitive Science*, 22, 25–51.

Benacerraf, P. (1965). What numbers could not be. In: P. Benacerraf and H. Putnam (1983), *Philosophy of Mathematics: Selected Readings* (2nd edn). Cambridge: Cambridge University Press.

Benacerraf, P. (1973). Mathematical truth. In: P. Benacerraf and H. Putnam (1983), *Philosophy of Mathematics: Selected Readings* (2nd edn). Cambridge: Cambridge University Press.

Bickhard, M. (1988). Piaget on variation and selection models: Structuralism, logical necessity, and interactivism. *Human Development*, 31, 274–312.

Bickhard, M. (2002). The biological emergence of representation. In: T. Brown and L. Smith (eds), *Reductionism and the Development of Knowledge*. Mahwah, NJ: Erlbaum.

Bickhard, M., & Terveen, L. (1995). *Foundational Issues in Artificial Intelligence and Cognitive Science*. Amsterdam: North-Holland.

Bideaud, J. (1992). *Pathways to Number*. Hillsdale, NJ: Erlbaum.

Black, P. (1998a). *Testing: Friend or Foe?* London: Falmer.

Black, P. (1998b). Learning, league tables and national assessment. *Oxford Review of Education*, 24, 57–68.

Black, P., & Wiliam, D. (1998). Assessment and classroom learning. *Assessment in Education*, *5*, 7–74.

Boden, M. (1979). *Piaget*. Brighton: Harvester Press.

Bond, T. (1997). Measuring development: Examples from Piaget's theory. In: L. Smith, J. Dockrell and P. Tomlinson (eds), *Piaget, Vygotsky, and Beyond*. London: Routledge.

Bond, T. (2001). Building a theory of formal operational thinking: Inhelder's psychology meets Piaget's epistemology. In: A. Tryphon and J. Vonèche (eds), *Working with Piaget: Essays in Honour of Bärbel Inhelder*. Hove: Psychology Press.

Boole, G. (1854/1958). *The Laws of Thought*. New York: Dover Publications.

Borowski, E., & Borwein, J. (1989). *Dictionary of Mathematics*. London: Collins.

Braine, M., & Rumain, B. (1983). Logical reasoning. In: P. Mussen (ed.), *Handbook of Child Psychology* (vol. 3). New York: Wiley.

Brainerd, C. (1973). Judgments and explanations as criteria for the presence of cognitive structures. *Psychological Bulletin*, *79*, 172–179.

Bringuier, J.-C. (1977/1980). *Conversations with Jean Piaget*. Chicago: University of Chicago Press.

Brown, A. (1997). Transforming schools into communities of thinking and learning about serious matters. *American Psychologist*, *52*, 399–413.

Brown, H. (1988). *Rationality*. London: Routledge.

Brown, T. (2001). Bärbel Inhelder and the fall of Valhalla. In: A. Tryphon and J. Vonèche (eds), *Working with Piaget: Essays in Honour of Bärbel Inhelder*. Hove: Psychology Press.

Braine, M., & O'Brien, D. (1998). *Mental Logic*. Mahwah, NJ: Erlbaum.

Bruner, J. (1960). *The Process of Education*. Cambridge, MA: Harvard University Press.

Bruner, J. (1966). *Towards a Theory of Education*. Cambridge, MA: Harvard University Press.

Bruner, J. (1996). *The Culture of Education*. Cambridge, MA: Harvard University Press.

Bryant, P. (1995). Children and arithmetic. *Journal of Child Psychology and Psychiatry*, *59*, 179–95.

Bryant, P. (1997). Piaget, mathematics, and Vygotsky. In: L. Smith, J. Dockrell and P. Tomlinson (eds), *Piaget, Vygotsky, and Beyond* (pp. 224–241). London: Routledge.

Bryant, P., & Kopytynska, H. (1976). Spontaneous measurement by young children. *Nature*, *260*, 773.

Burge, T. (1993). Content preservation. *The Philosophical Review*, *102*, 457–488.

Byrnes, J., & Duff, M. (1989). Young children's comprehension of modal expressions. *Cognitive Development*, *4*, 369–387.

Byrnes, J., & Beilin, H. (1991). The cognitive basis of uncertainty. *Human Development*, *34*, 189–203.

Campbell, D., & Stanley, J. (1963). *Experimental and Quasi-experimental Designs for Research*. Chicago: Rand McNally.

Carey, S. (1985). *Conceptual Change in Science*. Cambridge, MA: MIT Press.

Carey, S. (1999). Sources of conceptual change. In: E. Scholnick, K. Nelson, S. Gelman and P. Miller (eds), *Conceptual Development: Piaget's Legacy*. Mahwah, NJ: Erlbaum Associates.

Carnap, R. (1962). *Logical Foundations of Probability* (2nd edn). Chicago: University of Chicago Press.

Carruthers, P. (1996). *Language, Thought and Consciousness*. Cambridge: Cambridge University Press.

Case, R. (1985). *Intellectual Development*. New York: Academic Press.

Case, R. (1998). The development of conceptual structures. In: W. Damon (ed.), *Handbook of Child Psychology* (vol. 2, 5th edn). New York: Wiley.

Case, R. (1999). Conceptual development in the child and in the field: A personal view of the Piagetian legacy. In: E. Scholnick, K. Nelson, S. Gelman and P. Miller (eds), *Conceptual Development: Piaget's Legacy*. Mahwah, NJ: Erlbaum Associates.

Chandler, M. (1997). Stumping for progress in a post-modern world. In: E. Amsel and K. Renniger (eds), *Change and Development*. Mahwah, NJ: Erlbaum.

Chapman, M. (1988). *Constructive Evolution*. Cambridge: Cambridge University Press.

Chapman, M. (1992). Equilibration and the dialects of organization. In: H. Beilin and P. Pufall (eds), *Piaget's Theory: Prospects and Possibilities*. Hillsdale, NJ: Erlbaum.

Cheng, P., & Holyoak, K. (1989). On the natural selection of reasoning theories. *Cognition, 33*, 285–313.

Chinen, A. (1984). Modal logic: A new paradigm of development and late-life potential. *Human Development, 27*, 42–56.

Cohen, L. J. (1986). *The Dialogue of Reason*. Oxford: Oxford University Press.

Coles, M. (1993). Teaching thinking. *Educational Psychology, 13*, 333–344.

Cook, T., & Campbell, D. (1979). *Quasi-experimentation: Design and Analysis for Field Settings*. Chicago: Rand McNally.

Daintith, J., & Nelson, R. (1989). *Dictionary of Mathematics*. London: Penguin Books.

Damon, W. (1977). *The Social World of the Child*. San Francisco: Jossey-Bass.

Davies, R. B., & Pickles, A. R. (1985). A panel study of life cycle effects in residential mobility. *Geographical Analysis, 17*, 199–216.

Davydov, V. (1995). The influence of Vygotsky on education. *Educational Researcher, 24*(3), 12–21.

Dehaene, S. (1997). *The Number Sense*. London: Allen Lane.

Demetriou, A. (1998). Nooplasis: 10+1 postulates about the formation of mind. *Learning and Instruction, 8*, 271–287.

Demetriou, A., & Kazi, S. (2001). *Unity and Modularity in the Mind and the Self*. London: Routledge.

DES (1989). *From Policy to Practice*. National Curriculum. London: HMSO.

DES (1991). *Mathematics 5 16*. Department of Education and Science. London: HMSO.

Desforges, C. (1995). How does experience affect theoretical knowledge for teaching? *Learning and Instruction, 5*, 385–400.

Devries, R. (1987). *Programs of Early Education*. London: Longman.

Dewey, J. (1966). *Selected Educational Writings*. London: Heinemann.

DfEE (1997). *Excellence in Education*. White Paper. London: Stationery Office.

DfEE (1999). *Mathematics*. National Curriculum. Department for Education and Employment. London: The Stationery Office. [website: *www.nc.uk.net*]

Dillon, J. (1990). *The Practice of Questioning*. London: Routledge.

Donaldson, M. (1992). *Human Minds*. London: Allen Lane Press.

Douglis, A. (1970). *Ideas in Mathematics*. Philadelphia: W. B. Saunders Company.

Duckworth, E. (1996). *The Having of Wonderful Ideas* (2nd edn). New York: Teachers College Press.

Ducret, J.-J. (1994). *Jean Piaget: Savant et Philosophe* (2 vols). Genève: Droz.

Dummett, M. (1981). *Frege: Philosophy of Language* (2nd edn). London: Duckworth.

Dummett, M. (1991). *Frege: Philosophy of Mathematics*. London: Duckworth.

Elkind, D. (1967). Piaget's conservation problems. *Child Development, 38*, 15–27.

Erickson, F. (1992). Why the clinical trial doesn't work. *Educational Researcher, 21*(5), 9–11.

Falmagne, R. J., & Gonsalves, J. (1995). Deductive inference. *Annual Review of Psychology, 46*, 525–559.

Feldman, D. (1995). Learning and development in nonuniversal theory. *Human Development*, *38*, 315–321.

Fenstermacher, G. (1986). Philosophy of research on teaching: three aspects. In: M. Wittrock (ed.), *Handbook of Research on Teaching* (3rd edn). New York: Macmillan.

Feuerstein, R. (1979). *Dynamic Assessment of Retarded Performance*. Baltimore: University Parks Press.

Feuerstein, R. (1980). *Instrumental Enrichment*. Baltimore: University Parks Press.

Fiddick, L., Cosmides, L., & Tooby, J. (2000). No interpretation without representation: The role of domain-specific representations and inferences in the Wason selection task. *Cognition*, *77*, 1–79.

Field, H. (1980). *Science Without Numbers*. Oxford: Blackwell.

Flavell, J. (1970). Concept development. In: P. Mussen (ed.), *Carmichael's Manual of Child Psychology* (vol. 1) (3rd edn). New York: Wiley.

Flavell, J. (1982). On cognitive development. *Child Development*, *53*, 1–10.

Flavell, J., Miller, P., & Miller, S. (1993). *Cognitive Development* (3rd edn). Englewood Cliffs, NJ: Prentice Hall.

Fodor, J. (1975). *The Language of Thought*. Cambridge, MA: Harvard University Press.

Foltz, C., Overton, W., & Ricco, R. (1995). Proof construction. Reprinted in: L. Smith (ed.) (1996), *Critical Readings on Piaget*. London: Routledge.

Frege, G. (1879/1972). *Conceptual Notation and Related Articles*. Oxford: Clarendon Press.

Frege, G. (1881/1979). Boole's logical calculus and the *Concept-script*. In: G. Frege (ed.), *Posthumous Papers* (pp. 9–46). Oxford: Blackwell.

Frege, G. (1884/1950). *The Foundations of Arithmetic*. Oxford: Blackwell.

Frege, G. (1891/1979). Logic. In: G. Frege (ed.), *Posthumous Papers*. Oxford: Blackwell.

Frege, G. (1892/1960). On sense and meaning. In: P. Geach and M. Black (eds), *Translations from the Philosophical Writings of Gottlob Frege* (3rd edn). Oxford: Blackwell.

Frege, G. (1897/1979). Logic. In: G. Frege (ed.), *Posthumous Papers*. Oxford: Blackwell.

Frege, G. (1903/1979). Logical defects in mathematics. In: G. Frege (ed.), *Posthumous Papers*. Oxford: Blackwell.

Frege, G. (1906/1979). Key sentences on logic. In: G. Frege (ed.), *Posthumous Papers*. Oxford: Blackwell.

Frege, G. (1915/1979). My basic logical insights. In: G. Frege (ed.), *Posthumous Papers*. Oxford: Blackwell.

Frege, G. (1918/1977). Thoughts. In: G. Frege (ed.), *Logical Investigations*. Oxford: Blackwell.

Frege, G. (1980). *Philosophical and Mathematical Correspondence*. Oxford: Blackwell.

Fuson, K. *et al.* (1988). Effects of collection terms on class-inclusion and on number tasks. *Cognitive Psychology*, *20*, 96–120.

Galton, M. (1999). *Inside the Primary Classroom 20 Years On*. London: Routledge.

Garnham, A., & Oakhill, J. (1994). *Thinking and Reasoning*. Oxford: Blackwell.

Gelman, R. (1972). Logical capacity of very young children: Number invariance rules. In: L. Smith (ed.) (1992), *Jean Piaget: Critical Assessments* (vol. 3). London: Routledge.

Gelman, R. (1978). Cognitive development. *Annual Review of Psychology*, *29*, 297–332.

Gelman, R. (1997). Constructing and using conceptual competence. *Cognitive Development*, *12*, 305–313.

Gelman, R., & Gallistel, R. (1978). *The Child's Understanding of Number*. Cambridge, MA: Harvard University Press.

Gelman, R., Meck, E., & Merkin, S. (1986). Young children's numerical competence. *Cognitive Development*, *1*, 1–9.

Gelman, R., & Williams, E. (1998). Enabling constraints for cognitive development and learn-
ing. In: W. Damon (ed.), *Handbook of Child Psychology* (5th edn). New York: Wiley.
Gigerenzer, G. (1989). *The Empire of Chance: How Probability Changed Science and Everyday
Life*. Cambridge: Cambridge University Press.
Gigerenzer, G. (1993). The superego, the ego, and the id of statistical reasoning. In: G. Keren
and C. Lewis (eds), *A Handbook for Data Analysis in the Behavioural Sciences: Methodo-
logical Issues*. Hillsdale, NJ: Erlbaum Associates.
Gigerenzer, G. (1998). Psychological challenges for normative models. In: D. Gabbay and
P. Smets (eds), *Handbook of Defensible Reasoning and Uncertainty Management Systems*
(vol. 1). Amsterdam: Kluwer Academic.
Ginsburg, H. (1981). Piaget and education: The contribution and limits of genetic epistemology.
In: I. Sigel, D. Brodzinsky and R. Golinkoff (eds), *New Directions in Piagetian Theory and
Practice*. Hillsdale, NJ: Erlbaum.
Ginsburg, H. (1997). *Entering the Child's Mind*. Cambridge: Cambridge University Press.
Goblot, E. (1929). *Traité de Logique*. Paris: Colin.
Goldman, A. (1986). *Epistemology and Cognition*. Cambridge, MA: Harvard University Press.
Goldstein, H. (1991) *Assessment in schools*. London: Institute for Public Policy
Goodman, N. (1979). *Fact, Fiction, and Forecast* (3rd edn). Hassocks: Harvester Press.
Gréco, P. (1963). Le progrès des inférences itératives et des notions arithmétiques chez
l'enfant et l'adolescent. In: P. Gréco, B. Inhelder, B. Matalon and J. Piaget (eds), *La Forma-
tion des Raisonnements Récurrentiels*. Paris: Presses Universitaires de France.
Grice, P. (1989). *Studies in the Way of Words*. Cambridge, MA: Harvard University Press.
Haack, S. (1978). *Philosophy of Logics*. Cambridge: Cambridge University Press.
Hall, E. (1970). A conservation with Jean Piaget and Bärbel Inhelder. *Psychology Today, 3*,
25–32.
Hamilton, D., & Parlett, M. (1977). *Beyond the Numbers Game*. London: Macmillan.
Harris, P. (1986). *Designing and Reporting Experiments*. Buckingham: Open University Press.
Hartnett, P., & Gelman, R. (1997). Early understanding of numbers. *Learning and Instruction,
8*, 341–374.
Hawkins, J., Pea, R., Glick, J., & Scribner, S. (1984). "Merds that laugh don't like mushrooms":
Evidence for deductive reasoning by preschoolers *Developmental Psychology, 20*, 584–594.
Heath, T. (1956). *The Thirteen Books of Euclid's Elements*. New York: Dover.
Heckman, J. J. (1981). Statistical models for discrete panel data. In: C. F. Manski and
D. McFadden (eds), *Structural Analysis of Discrete Data with Econometric Applications*
(pp. 114–178). Cambridge, MA: MIT Press.
Hersh, R. (1997). *What is Mathematics Really?* New York: Cape.
Hintikka, J. (1996). *The Principles of Mathematics Revisited*. Cambridge: Cambridge University
Press.
Hodges, A. (1983). *Alan Turing: the Enigma of Intelligence*. London: Unwin.
Hofer, B., & Pintrich, P. (1997). The development of epistemological theories: Beliefs about
knowledge and knowing and their relation to learning. *Review of Educational Research, 67*,
88–140.
Holyoak, K., & Spellman, B. (1993). Thinking. *Annual Review of Psychology, 44*, 265–315.
Holzman, L. (1997). *schools for Growth: Radical Alternatives to Current Educational Models*.
Mahwah, NJ: Erlbaum.
Hughes, M. (1986). *Children and Number*. Oxford: Blackwell.
Hughes, M. (2000). *Numeracy and Beyond*. Buckingham: Open University Press.
Inhelder, B. (1954/2001). The experimental approach of children and adolescents. In: A. Tryphon
and J. Vonèche (eds), *Working with Piaget: Essays in Honour of Bärbel Inhelder*. Hove:
Psychology Press.

Inhelder, B. (1956). Criteria of the stages of development. In: J. Tanner and B. Inhelder (eds), *Discussions on Child Development* (vol. 1). London: Tavistock.

Inhelder, B., & Piaget, J. (1955/1958). *The Growth of Logical Thinking.* London: Routledge & Kegan Paul.

Inhelder, B., & Piaget, J. (1963). Itération et récurrence. In: P. Gréco, B. Inhelder, B. Matalon and J. Piaget (eds), *La Formation des Raisonnements Récurrentiels.* Paris: Presses Universitaires de France.

Inhelder, B., & Piaget, J. (1979/1980). Procedures and structures. In: D. Olson (ed.), *The Social Foundations of Language.* New York: Norton.

Inhelder, B., Sinclair, H., & Bovet, M. (1974). *Learning and the Development of Cognition.* London: Routledge & Kegan Paul.

Isaacs, N. (1951). Critical notice: *Traité de logique. British Journal of Psychology, 42*, 185–188.

Johnson, C., & Harris, P. (1994). Magic: Special but not excluded. *British Journal of Developmental Psychology, 12*, 35–51.

Johnson-Laird, P. (1983). *Mental Models.* Cambridge: Cambridge University Press.

Johnson-Laird, P. (1999). Deductive reasoning. *Annual Review of Psychology, 50*, 109–135.

Johnson-Laird, P., & Savary, F. (1999). Illusory inferences: A novel class of erroneous deductions. *Cognition, 71*, 191–229.

Kamii, C. (1982). Encouraging thinking in mathematics. Reprinted in: L. Smith (ed.) (1992), *Jean Piaget: Critical Assessments* (vol. 3). London: Routledge.

Kant, I. (1933). *Critique of Pure Reason* (2nd edn). London: Macmillan.

Kant, I. (1948). *Groundwork of the Metaphysics of Morals.* In: H. Paton (ed.), *The Moral Law.* London: Hutchinson.

Kargopoulos, P., & Demetriou, A. (1998). Logical and psychological partitioning of the mind: Depicting the same map? *New Ideas in Psychology, 16*, 61–87.

Karmiloff-Smith, A. (1994). Précis of *Beyond Modularity: A Developmental Perspective on Cognitive Science. Behavioural and Brain Sciences, 17*, 693–745.

Katz, J. (1995). What mathematical knowledge could be. *Mind, 104*, 491–522.

Keil, F., & Lockhart, K. (1999). Explanatory understanding in conceptual development. In: E. Scholnick, K. Nelson, S. Gelman and P. Miller (eds), *Conceptual Development: Piaget's Legacy.* Mahwah, NJ: Erlbaum Associates.

Kenny, A. (1995). *Frege.* London: Penguin Books.

Keppel, G. (1991). *Design and Analysis: A Researcher's Handbook* (3rd edn). Engelwood Cliffs, NJ: Prentice Hall.

Kinnear, P., & Gray, C. (1997). *SPSS for Windows.* Hove: Psychology Press.

Kitchener, R. (1986). *Piaget's Theory of Knowledge.* New Haven, CT: Yale University Press.

Kitchener, R. (1991). Jean Piaget: The unknown sociologist? *British Journal of Sociology, 42*, 421–442.

Kitcher, P. (1983). *The Nature of Mathematical Knowledge.* New York: Oxford University Press.

Klahr, D. (1999). The conceptual habitat: In what kind of system can concepts develop? In: E. Scholnick, K. Nelson, S. Gelman and P. Miller (eds), *Conceptual Development: Piaget's Legacy.* Mahwah, NJ: Erlbaum Associates.

Kneale, W., & Kneale, M. (1962). *The Development of Logic.* Oxford: Oxford University Press.

Kohlberg, L. (1987). *Child Psychology and Childhood Education.* London: Longman.

Kornblith, H. (1985). *Naturalizing Epistemology.* Cambridge, MA: MIT Press.

Kozulin, A. (1998). *Psychological Tools.* Cambridge, MA: Harvard University Press.

Kripke, S. (1980). *Naming and Necessity.* Oxford: Blackwell.

Kuhn, D. (1979). The application of Piaget's theory of cognitive development to education. *Harvard Educational Review*, *49*, 340–360.

Kuhn, D. (1989). Children and adults as intuitive scientists. *Psychological Review*, *96*, 674–689.

Kuhn, D. (1991). *The Skills of Argument*. Cambridge: Cambridge University Press.

Kuhn, D. (1997). The view from giants' shoulders. In: L. Smith, J. Dockrell and P. Tomlinson (eds), *Piaget, Vygotsky and Beyond*. London: Routledge.

Kuhn, D., Amsel, E., & O'Loughlin, M. (1988). *The Development of Scientific Thinking Skills*, Orlando, FL: Academic Press.

Kuhn, D., Schauble, L., & Garcia-Mila, M. (1992). Cross-domain development of scientific reasoning. *Cognition and Instruction*, *9*, 285–327.

Kukla, A. (1989). Nonempirical issues in psychology. *American Psychologist*, *44*, 785–794.

Lakatos, I. (1976). *Proofs and Refutations: The Logic of Mathematical Discovery*. Cambridge: Cambridge University Press.

Langer, J. (1980). *The Origins of Logic: Six to Twelve Months*. New York: Academic Press.

Langer, J. (1998). Phylogenetic and ontogenetic origins of cognitive classification. In: J. Langer and M. Killen (eds), *Piaget, Evolution and Development*. Mahwah, NJ: Erlbaum Associates.

Larochelle, M. *et al.* (1998). *Constructivism and Education*. Cambridge: Cambridge University Press.

Laudan, L. (1984). *Science and Values*. Berkeley, CA: University of California Press.

Leach, J. (1999). Students' understanding of the co-ordination of theory and evidence in science. *International Journal of Science Education*, *21*, 789–806.

Leat, D. (1999). Rolling the stone uphill: Teacher development and the implementation of Thinking Skills programmes. *Oxford Review of Education*, *25*, 387–403.

Leibniz, G. W. (1981). *New Essays on Human Understanding*. Cambridge: Cambridge University Press.

Lemmon, E. J. (1966). *Beginning Logic*. London: Nelson.

Lewis, C., & Langford, C. (1959). *Symbolic Logic*. New York: Dover Publications.

Lewis, D. (1986). *On the Plurality of Possible Worlds*. Oxford: Blackwell.

Light, R., & Pillemer, D. (1982). Numbers and narrative: Combining their strengths in research reviews. *Harvard Educational Review*, *52*, 1–26.

Lourenço, O., & Machado, A. (1996). In defence of Piaget's theory: A reply to 10 common criticisms. *Psychological Review*, *103*, 143–164.

Machado, A., Lourenço, O., & Silva, F. (2000). Facts, concepts and theories: The shape of psychology's epistemic triangle. *Behaviour and Philosophy*, *28*, 1–40.

Marcus, R. B. (1993). *Modalities: Philosophical Essays*. New York: Oxford University Press.

Markman, E. (1978). Empirical versus logical solutions to part–whole comparison problems concerning classes and collections. *Child Development*, *49*, 168–177.

Markovits, H. (1993). The development of conditional reasoning. *Merrill Palmer Quarterly*, *39*, 133–160.

Markovits, H. *et al.* (1996). Reasoning in young children: Fantasy and information retrieval. *Child Development*, *67*, 2857–2872.

Marton, F., & Booth, S. (1997). *Learning and Awareness*. Mahwah, NJ: Erlbaum Associates.

Mason, J. (1996). *Qualitative Researching*. London: Sage.

Massy, W. F., Montgomery, D. M., & Morrison, D. G. (1970). *Stochastic Models for Buying Behaviour*. Cambridge, MA: MIT Press.

Maxwell, J. (1992). Understanding and validity in qualitative research. *Harvard Educational Review*, *62*, 279–300.

Mays, W. (1992). Piaget's logic and its critics: A deconstruction. *Archives de Psychologie*, *60*, 45–70.

Mays, W. (2000). Piaget's sociology revisited. *New Ideas in Psychology*, *18*, 241–260.

Mays, W., & Smith, L. (2002). Harré on Piaget's *Sociological Studies*. *New Ideas in Psychology*, 20.

McCullagh, P., & Nelder, J. A. (1989). *Generalized Linear Models*. London: Chapman & Hall.

McGinn, C. (1991). *The Problem of Consciousness: Essay Toward a Resolution*. Oxford: Blackwell.

McGuinness, C., & Nisbet, J. (1991). Teaching thinking in Europe. *British Journal of Educational Psychology*, *61*, 174–186.

Mercer, N. (1995). *The Guided Construction of Knowledge*. London: Multilingual Matters.

Michell, J. (1997). Quantitative science and the definition of *measurement* in psychology. *British Journal of Psychology*, *88*, 355–383.

Miller, P., & Coyle, T. (1999). Developmental change: Lessons from microgenesis. In: E. Scholnick, K. Nelson, S. Gelman and P. Miller (eds), *Conceptual Development: Piaget's Legacy*. Mahwah, NJ: Erlbaum Associates.

Miller, S. (1986). Certainty and necessity in the understanding of Piagetian concepts. *Developmental Psychology*, *22*, 3–18.

Mooij, J. (1966). *La Philosophie des Mathématiques de Henri Poincaré*. Paris: Gauthier-Villars.

Moore, R. (1994). Making the transition to formal proof. *Educational Studies in Mathematics*, *27*, 249–266.

Morris, A., & Sloutsky, V. (1998). Understanding of logical necessity. *Child Development*, *69*, 721–741.

Moser, P. (1995). Epistemology. In: L. R. Audi (ed.), *The Cambridge Dictionary of Philosophy* (2nd edn). Cambridge: Cambridge University Press.

Moshman, D. (1990). The development of metalogical understanding. In: W. Overton (ed.), *Reasoning, Necessity and Logic*. Hillsdale, NJ: Erlbaum.

Moshman, D. (1994). Reason, reasons and reasoning. *Theory and Psychology*, *4*, 245–260.

Moshman, D. (1998). Cognitive development beyond childhood. In: L. W. Damon (ed.), *Handbook of Child Psychology* (vol. 2) (5th edn). New York: Wiley.

Moshman, D., & Franks, B. (1986). Development of the concept of inferential validity. *Child Development*, *57*, 153–165.

Moshman, D., & Timmons, M. (1982). The construction of logical necessity. *Human Development*, *25*, 309–323.

Müller, U., Sokol, B., & Overton, W. (1998). Development of mental representation. *Developmental Review*, *19*, 155–201.

Müller, U., Sokol, B., & Overton, W. (1999). Developmental sequences in class reasoning and prepositional reasoning. *Journal of Experimental Child Psychology*, *74*, 69–106.

Murray, F. (1981). The conservation paradigm: Conservation of conservation research. In: I. Sigel, D. Brodzinsky and R. Golinkoff (eds), *New Directions to Piagetian Theory and Practice*. Hillsdale, NJ: Erlbaum.

Murray, F. (1990). The conversion of truth into necessity. In: W. Overton (ed.), *Reasoning Necessity and Logic*. Hillsdale, NJ: Erlbaum.

Nagel, E. (1961). *The Structure of Science*. New York: Routledge & Kegan Paul.

Nagel, T. (1997). *The Last Word*. New York: Oxford University Press.

Nickerson, R. (1985). *The Teaching of Thinking*. Hillsdale, NJ: Erlbaum.

Nunes, T., & Bryant, P. (1996). *Children Doing Mathematics*. Oxford: Blackwell.

Nunes, T., & Bryant, P. (1997). *Learning and Teaching Mathematics*. Hove: Psychology Press.

Oden, G. (1987). Concept, knowledge and thought. *Annual Review of Psychology*, *38*, 203–228.

Parrat-Dayan, S. (1993). La réception de l'oeuvre de Piaget dans le milieu pédagogique des années 1920–1930. *Revue Francaise de Pédagogie, 104*, 73–83.

Pareto, W. (1963). *A Treatise on General Sociology*. New York: Dover.

Parsons, C. (1960). Critical notice. *British Journal of Psychology, 51*, 75–84.

Piaget, J. (1918). *Recherche*. Lausanne: La Concorde.

Piaget, J. (1923). La psychologie des valeurs religieuses. In: Association Chrétienne d'Etudiants de la Suisse Romande (ed.), *Sainte-Croix 1922*, 38–82.

Piaget, J. (1924/1928). *Judgment and Reasoning in the Child*. London: Routledge & Kegan Paul.

Piaget, J. (1926/1929). *The Child's Conception of the World*. London: Routledge & Kegan Paul.

Piaget, J. (1932/1932). *The Moral Judgment of the Child*. London: Routledge & Kegan Paul.

Piaget, J. (1936/1953). *The Origins of Intelligence in the Child*. London: Routledge & Kegan Paul.

Piaget, J. (1937/1954). *The Construction of Reality in the Child*. London: Routledge & Kegan Paul.

Piaget, J. (1941). Le mécanisme du développement mental et les lois du groupement des opérations. *Archives de Psychologie, 28*, 215–285.

Piaget, J. (1942). *Classes, Relations, Nombres: Essai sur les Groupement de la Logistique et sur la Réversibilité de la Pensée*. Paris: Vrin.

Piaget, J. (1945/1962). *Play, Dreams and Imitation in Childhood*. London: Routledge & Kegan Paul.

Piaget, J. (1947). Avant-Propos de la troisième édition. *Le Jugement et le Raisonnement chez l'Enfant*. Neuchatel: Delachaux et Niestlé.

Piaget, J. (1949). *Traité de Logique*. Paris: Colin.

Piaget, J. (1950). *Introduction à L'épistémologie Génétique. (Vol. 1.) La Pensée Mathématique*. Paris: Presses Universitaires de France.

Piaget, J. (1952). Autobiography. In: E. Boring (ed.), *A History of Psychology in Autobiography*, vol. 4. Worcester, MA: Clark University Press.

Piaget, J. (1953). *Logic and Psychology*. Manchester: Manchester University Press.

Piaget, J. (1960). The general problems of the psychobiological development of the child. In: J. Tanner and D. Inhelder (eds), *Discussions on Child Development* (vol. 4). London: Tavistock.

Piaget, J. (1961/1966). Part II. In: E. Beth and J. Piaget (eds), *Mathematical Epistemology and Psychology*. Dordrecht: Reidel.

Piaget, J. (1962/1995). Commentary on Vygotsky's criticisms. *New Ideas in Psychology, 13*, 325–40. Reprinted as Piaget, J. (2000). Commentary on Vygotsky's criticisms. *New Ideas in Psychology, 18*, 241–259.

Piaget, J. (1967a). Les méthodes de l'épistémologie. In: J. Piaget (ed.), *Logique et Connaissance Scientifique*. Paris: Gallimard.

Piaget, J. (1967b). Epistémologie de la logique. In: J. Piaget (ed.), *Logique et Connaissance Scientifique*. Paris: Gallimard.

Piaget, J. (1967/1971). *Biology and Knowledge*. Edinburgh: Edinburgh University Press.

Piaget, J. (1968). *On the Development of Memory and Identity*. Barre, MA: Clark University Press.

Piaget, J. (1969/1971). *Science of Education and the Psychology of the Child*. London: Longman.

Piaget, J. (1970/1977). *Psychology and Epistemology*. London: Penguin.

Piaget, J. (1970/1983). Piaget's theory. In: P. Mussen (ed.) (1983), *Handbook of Child Psychology*, 4th edn New York: Wiley.

Piaget, J. (1973). Comments on mathematical education. In: A. Howson (ed.), *Developments in Mathematical Education*. Cambridge: Cambridge University Press.

Piaget, J. (1974/1976). *Success and Understanding*. London: Routledge & Kegan Paul.

Piaget, J. (1974/1977). *The Grasp of Consciousness*. London: Routledge & Kegan Paul.

Piaget, J. (1975/1985). *Equilibration of Cognitive Structures*. Chicago: University of Chicago Press.

Piaget, J. (1977). The role of action in the development of thinking. In: W. Overton and J. M. Gallagher (eds), *Knowledge and Development* (vol. 1). New York: Plenum Press.

Piaget, J. (1977/1980). Twelfth conversation: Concerning creativity. In: J.-C. Bringuier (ed.), *Conversations with Jean Piaget*. Chicago: University of Chicago Press.

Piaget, J. (1977/1986). Essay on necessity. *Human Development, 29*, 301–314.

Piaget, J. (1977/1995). *Sociological Studies*. London: Routledge.

Piaget, J. (1977/2000). *Studies in Reflective Abstraction*. Hove: Psychology Press.

Piaget, J. (1978). *Recherches sur la Généralisation*. Paris: Presses Universitaires de France.

Piaget, J. (1983/1987). *Possibility and Necessity* (vol. 2). Minneapolis: University of Minnesota Press.

Piaget, J. (1998). *De la Pédagogie*. Paris: Odile Jacob.

Piaget, J., & Garcia, J. (1987/1991). *Toward a Logic of Meanings*. Hillsdale, NJ: Erlbaum Associates.

Piaget, J., & Inhelder, B. (1941/1974). *The Child's Construction of Quantities*. London: Routledge & Kegan Paul.

Piaget, J., & Inhelder, B. (1961). Introduction à la séconde édition. *Le Développement des Quantités Physiques chez l'Enfant*. Neuchatel: Delachaux et Niestlé.

Piaget, J., & Inhelder, B. (1966/1971). *Mental Imagery in the Child*. London: Routledge & Kegan Paul.

Piaget, J., Inhelder, B., & Szeminska, A. (1948/1960). *Child's Conception of Geometry*. London: Routledge & Kegan Paul.

Piaget, J., & Szeminska, A. (1941/1952). *The Child's Conception of Number*. London: Routledge & Kegan Paul.

Piaget, J., & Voyat, G. (1968). Recherche sur l'identité d'un corps en développement et sur celle du movement transitif. In: J. Piaget, H. Sinclair and Vinh Bang (eds), *Epistémologie et Psychologie de l'Identité* (pp. 1–82). Paris: Presses Universitaires de France.

Piéraut-Le Bonniec, G. (1980). *The Development of Modal Reasoning*. New York: Academic Press.

Phillips, D. C. (1997). How, why, what, when, and where: Perspectives on constructivism in psychology and education. *Issues in Education, 3*, 151–194, 273–284.

Plato (1941). *Republic*. Oxford: Oxford University Press.

Poincaré, H. (1902/1905). *Science and Hypothesis*. London: The Walter Scott Publishing Co.

Poincaré, H. (1908/1952). *Science and Method*. New York: Dover Publications Inc.

Popper, K. (1963). *Conjectures and Refutations*. London: Routledge & Kegan Paul.

Popper, K. (1979). *Objective knowledge* (2nd edn). Oxford: Oxford University Press.

Porpodas, C. (1987). The one-question conservation experiment reconsidered. *Journal of Child Psychology and Psychiatry, 28*, 343–349.

Quine, W. (1972). *Methods of Logic* (3rd edn). London: Routledge & Kegan Paul.

Raven, J. (1994). *Managing Education for Effective schooling*. Oxford: Oxford Psychologists Press.

Reichenbach, H. (1947). *Elements of Symbolic Logic*. Toronto: Macmillan.

Resnick, L. (1987). *Education and Thinking*. Washington: National Academy Press.

Reynolds, D., & Farrell, S. (1996). *Worlds Apart*. Ofsted Report. London: The Stationery Office.

Ricco, R. (1993). Revising the logic of operations as a relevance logic: From hypothesis testing to explanation. *Human Development, 36*, 125–146.

Ricco, R. (1997). The development of proof construction in middle childhood. *Journal of Experimental Child Psychology, 66*, 279–310.

Rose, S., & Blank, M. (1974). The potency of context in children's cognition: An illustration through conservation. *Child Development, 45*, 499–502.

Ross, A. (1968). *Norms and Directives*. London: Routledge & Kegan Paul.

Ruffman, T., Perner, J., Olson, D., & Doherty, M. (1993). Reflecting on scientific thinking: Children's understanding of the hypothesis–evidence relation. *Child Development, 64*, 1617–1636.

Russell, B. (1903/1964). *The Principles of Mathematics* (2nd edn). London: George Allen & Unwin.

Russell, B. (1905/1992). On denoting. In: R. Marsh (ed.) (1992), *Logic and Knowledge*. London: Routledge.

Russell, B. (1912/1959). *The Problems of Philosophy*. London: Oxford University Press.

Russell, B. (1919). *Introduction to Mathematical Philosophy*. London: George Allen & Unwin.

Russell, B. (1959). *My Philosophical Development*. London: George Allen & Unwin.

Russell, J. (1983). Children's ability to discriminate between types of proposition. *British Journal of Developmental Psychology, 1*, 259–268.

Saussure, F. de (1960). *Course in General Linguistics*. London: Peter Owen.

Sainsbury, M. (1991). *Logical Forms*. Oxford: Blackwell.

Satterly, D. (1989). *Assessment in schools*. Oxford: Blackwell.

Scholnick, E. (1999). Piaget's legacy: Heirs to the house that Jean built. In: E. Scholnick, K. Nelson, S. Gelman and P. Miller (eds), *Conceptual Development: Piaget's Legacy*. Mahwah, NJ: Erlbaum.

Scholnick, E., & Wing, C. (1995). Logic in conversation: Comparative studies of deduction in children and adults. *Cognitive Development, 10*, 319–345.

Schrag, F. (1992) In defence of positivist research paradigms. *Educational Researcher, 21*(5), 5–8.

Searle, J. (1995). *The Construction of Social Reality*. New York: The Free Press.

Shayer, M. (1997). Piaget and Vygotsky: A necessary marriage for effective intervention. In: L. Smith, J. Dockrell and P. Tomlinson (eds), *Piaget, Vygotsky and Beyond*. London: Routledge.

Shayer, M. (1999). Cognitive acceleration through science education II: Its effects and scope. *International Journal of Science Education, 21*, 883–902.

Shayer, M., & Adey, P. (1981) *Towards a Science of Science Teaching*. London: Heinemann.

Shayer, M., Demetriou, A., & Pervez, M. (1988). The structure and scaling of concrete operational thought: Three studies in four countries. *Genetic, Social and General Psychology Monographs, 114*, 309–375.

Shephard, L. (2000). The role of assessment in a learning culture. *Educational Researcher, 29*(7), 4–14.

Siegal, M. (1997). *Knowing Children: Experiments in Conversation and Cognition* (2nd edn). Hove: Psychology Press.

Siegal, M. (1999). Language and thought: The fundamental significance of conversational awareness for cognitive development. *Developmental Science, 2*, 1–34.

Siegler, R. (1978). *Children's Thinking: What Develops?* Hillsdale, NJ: Erlbaum Associates.

Siegler, R. (1996). *The Emergence of Mind*. New York: Oxford University Press.

Simon, M. (1996). Beyond inductive and deductive reasoning: The search for a sense of knowing. *Educational Studies in Mathematics, 30*, 197–210.

Singh, S. (1997). *Fermat's Last Theorem.* London: Fourth Estate.

Skyrms, B. (1995). Induction. In: R. Audi (ed.), *The Cambridge Dictionary of Philosophy.* Cambridge: Cambridge University Press.

Smedslund, J. (1966). Les origines sociales de la décentration. In: F. Bresson and M. de Montmollen (eds), *Psychologie et Épistémologies Génétiques: Themes Piagétiens.* Paris: Dunod.

Smedslund, J. (1994). What kind of propositions are set forth in developmental research? Five case studies. *Human Development, 37,* 280–292.

Smith, L. (1987). A constructivist interpretation of formal operations. *Human Development, 30,* 341–354.

Smith, L. (1992). *Jean Piaget: Critical Assessments* (4 vols). London: Routledge.

Smith, L. (1993). *Necessary Knowledge.* Hove: Erlbaum Associates Ltd.

Smith, L. (1994). Reasoning models and intellectual development. In: A. Demetriou and A. Efklides (eds), *Mind, Reasoning and Intelligence.* Amsterdam: North Holland.

Smith, L. (1996a). Piaget's epistemology: Psychological and educational assessment. In: L. Smith (ed.), *Critical Readings on Piaget.* London: Routledge.

Smith, L. (1996b). With knowledge in mind: Novel transformation of the learner or transformation of novel knowledge. *Human Development, 39,* 257–263.

Smith, L. (1997). Necessary knowledge and its assessment in intellectual development. In: L. Smith, J. Dockrell, and P. Tomlinson (eds), *Piaget, Vygotsky and Beyond.* London: Routledge.

Smith, L. (1998a). On the development of mental representation. *Developmental Review, 18,* 202–227.

Smith, L. (1998b). Modal knowledge and maps of the mind. *New Ideas in Psychology, 16,* 115–124.

Smith, L. (1999a). What Piaget learned from Frege. *Developmental Review, 19,* 133–153.

Smith, L. (1999b). Epistemological principles for developmental psychology in Frege and Piaget. *New Ideas in Psychology, 17,* 83–118.

Smith, L. (1999c). Eight good questions for developmental epistemology and psychology. *New Ideas in Psychology, 17,* 137–148.

Smith, L. (1999d). Necessary knowledge in number conservation. *Developmental Science, 2,* 23–27.

Smith, L. (1999e). What exactly is constructivism in education? *Studies in Science Education, 33,* 149–160.

Smith, L. (1999f). Representation and knowledge are not the same thing. *Behavioural and Brain Sciences, 22,* 784–785.

Smith, L. (2000a). Children's reasoning, mathematical induction and developmental epistemology. In: A. Gagatsis, C. Constantinou and L. Kyriakides (eds), *Learning and Assessment in Mathematics and Science.* Nicosia: University of Cyprus.

Smith, L. (2000b). Deontic reasoning about the rules of social action. Paper presented at the Jean Piaget Society Annual Meeting, Montréal.

Smith, L. (2002a). Piaget's model. In: U. Goswami (ed.), *Handbook of Cognitive Development.* Oxford: Blackwell.

Smith, L. (2002b). From epistemology to psychology in the development of knowledge. In: T. Brown and L. Smith (eds), *Reductionism and the Development of Knowledge.* Mahwah, NJ: Erlbaum.

Smith, L. (2002c). Jean Piaget. In: J. Palmer (ed.), *100 Great Thinkers on Education.* London: Routledge.

Smith, L. (2002d). Developmental epistemology and education. In: J. Carpendale and U. Müller (eds), *Social Interaction and the Development of Knowledge*. Mahwah, NJ: Erlbaum.

Smith, L., Dockrell, J., & Tomlinson, P. (1997). Introduction. In: L. Smith, J. Dockrell and P. Tomlinson (eds), *Piaget, Vygotsky and Beyond*. London: Routledge.

Sophian, C. (1995). Representation and reasoning in early numerical development. *Child Development, 66*, 559–577.

Steffe, L., & Gale, J. (1995). *Constructivism in Education*. Hillsdale, NJ: Erlbaum.

Strawson, P. (1952). *Introduction to Logical Theory*. London: Methuen.

Stroud, B. (1979). Inference, belief, understanding. *Mind, 88*, 179–196.

Torrance, H., & Pryor, J. (1998). *Investigating Formative Assessment*. Milton Keynes: Open University Press.

Turiel, E. (1983). *The Development of Social Knowledge*. Cambridge: Cambridge University Press.

Tversky, A., & Kahneman, D. (1974). Judgment under uncertainty: Heuristics and biases. *Science, 185*, 1124–1131.

Vidal, F. (1994). *Piaget Before Piaget*. Cambridge, MA: Harvard University Press.

Vonèche, J.-J. (1999). The origins of Piaget's ideas about genesis and development. In: E. Scholnick, K. Nelson, S. Gelman and P. Miller (eds), *Conceptual Development: Piaget's Legacy*. Mahwah, NJ: Erlbaum Associates.

Vonèche, J.-J. (2001). Bärbel Inhelder's contribution to psychology. In: A. Tryphon and J. Vonèche (eds), *Working with Piaget: Essays in Honour of Bärbel Inhelder*. Hove: Psychology Press.

von Wright, G. H. (1957). *Logical Studies*. London: Routledge & Kegan Paul.

von Wright, G. H. (1971). *Explanation and Understanding*. London: Routledge & Kegan Paul.

von Wright, G. H. (1983a). *Practical Reason*. Oxford: Blackwell.

von Wright, G. H. (1983b). *Philosophical Logic*. Oxford: Blackwell.

von Wright, G. H. (1984). *Truth, Knowledge, and Modality*. Oxford: Blackwell.

Vygotsky, L. (1986). *Thought and Language* (2nd edn). Cambridge, MA: MIT Press.

Vygotsky, L. (1994). *A Vygotsky Reader*. Oxford: Blackwell.

Waismann, F. (1951). *Introduction to Mathematical Thinking* New York: Frederick Ungar Publishing Co.

Wang, H. (1996). *A Logical Journey: From Gödel to Philosophy*. Cambridge, MA: MIT Press.

Weinert, F., & Schneider, W. (1999). *Individual Development from Three to Twelve*. Cambridge: Cambridge University Press.

Whitburn, J. (2000). *Strength in Numbers*. London: NIESR.

Whitehead, A., & Russell, B. (1910/1970). *Principia Mathematica to *56*. Cambridge: Cambridge University Press.

Wilkinson, L. (1999). Statistical methods in psychology journals. *American Psychologist, 54*, 594–604.

Williams, E., & Shuard, H. (1970). *Primary Mathematics Today*. London: Longman.

Wittgenstein, L. (1972). *Tractatus Logico-Philosophicus* (2nd edn). London: Routledge & Kegan Paul.

Wittgenstein, L. (1998). *Culture and Value*. Oxford: Blackwell.

Index